SUPPORTING PROFESSIONAL GIS MANAGEMENT:

A VISION FOR THE GIS MANAGEMENT INSTITUTE

Greg Babinski, MA, GISP

Make big plans; aim high in hope and work, remembering that a noble, logical diagram once recorded will never die

-Daniel Hudson Burnham

Supporting Professional GIS Management: A Vision for the GIS Management Institute

Orzel Books

Edmonds, Washington

Library of Congress Control Number 2018935177
ISBN 978-0-9986495-2-8

Printed by CreateSpace, An Amazon.com Company

Address inquiries to the author at:

Greg Babinski
Orzel Books
201 Third Ave. N., Suite 316
Edmonds, WA 98020 USA
gbabinski@gmail.com

TABLE OF CONTENTS

LIST OF FIGURES

PREFACE

Why have I authored this work about the GIS Management Institute? Geographic information systems (GIS) have become a proven technology-based spatial data management infrastructure that is critical for the effective operation of local, regional, and national governments, as well as for many businesses and other non-government entities.

GIS is barely 50 years old. Even a small local government jurisdiction may invest millions of dollars in its GIS and spend as much as one percent of its budget on annual GIS operations. However, the study and development of the best means to operate and manage a GIS is in its infancy.

The GIS Management Institute (GMI) Vision evolved from the Urban and Regional Information Systems Association (URISA) and its interest in the GIS Capability Maturity Model. I developed the original GIS Capability Maturity Model in 2008 and then tested its use by cities and counties in Washington and Oregon States. In 2009, with my cooperation, URISA adopted the GIS Capability Maturity Model that I had developed as its own.

In 2011 URISA established the GIS Management Institute Committee to support professional operation and management of GIS. It developed the Geospatial Management Competency Model in 2012 and a revised GIS Capability Maturity Model in 2013. The GMI Vision evolved with: a) development by a committee of experienced GIS managers and academics, b) monthly briefings and discussion with the URISA Board, and c) worldwide outreach efforts to describe, discuss, and refine the GMI Vision. The URISA Board approved the GMI Vision and Business Plan in 2013 by inclusion in the Association Bylaws. For two years development of the GMI Vision proceeded with URISA support and with the work of dozens of volunteers.

But in 2015, the then Board removed the GMI Vision from the URISA Bylaws and reduced its scope. This action was taken by the Board without consultation with the GMI Committee.

An organization is needed for professional GIS managers and for those entities that operate GIS to research, study, test, develop, and evolve best practices for GIS operation and management. Outside the GIS Management Institute, there is no current organization that claims or performs this mission.

This work was originally drafted as an article for the *URISA Journal* in 2016. It was submitted to URISA on May 12, 2016 and posted for public review and comment on the *URISA Journal* web page as 'Under Review' for approximately 11 months. When I submitted the article I also signed and submitted a standard URISA copyright release that restricted me from publishing the article elsewhere.

At some point in April 2017, the article disappeared from the URISA website without prior notification or explanation. When I inquired about this disappearance, the then editor of the *Journal* informed me that he had decided that it was '...more suited as a URISA monograph than a journal article.'[i] On April 17, 2017 I agreed to have my work published as a URISA monograph, along with the GIS Capability Maturity Model and the Geospatial Management Competency Model.

However, on June 10, 2017, I was informed by the *URISA Journal* editor that he and three other anonymous members of the URISA Professional and Educational

Material Committee who read my manuscript judged the 'tone' as unprofessional and that it contained factual errors. I was then informed that my work would not be published unless I agreed in advance to changes that an anonymous group of reviewers would provide to me. I was also refused permission by URISA to include the full GIS Capability Maturity Model and the Geospatial Management Competency Model in this work.[ii]

It was then that I decided to publish this monograph on my own. I requested examples of errors and unprofessional tone from the *Journal* editor, but I received no reply. I requested formal release from my signed URISA copyright release, but to date I have not received a reply.

I perceived then that URISA, at that point in time, was no longer fully committed to the GIS Management Institute that it had created and to the GMI Vision that it had previously adopted. I sincerely hope that future events prove this belief wrong. The Board action in 2015 to change the scope of the GMI is clearly within its power. However, I do not understand the basis for that decision.

URISA has an annual awards program that recognizes the contributions of its members to the Association. It has never seen fit to recognize the dozens of URISA members who devoted thousands of hours to develop the GMI and its components for the Association. The decision by URISA to not publish an uncensored description of the development of the GIS Management Institute and of the GMI Vision through 2016 indicates a refusal to even acknowledge the work of those who developed that vision.

The GIS Code of Ethics requires that GIS professionals respect the work of their colleagues.

I am indebted to all my colleagues who worked on the development of the GMI. Naturally, any errors in this work are solely my responsibility.

I authored this work to document the development of the GMI Vision, to recognize the seven years of work of many dedicated GIS professionals, and so that the GMI Vision can be readopted in the future by URISA – or by some other organization - should it choose to do so.

INTRODUCTION

The purpose of this work is to document the development of the GIS Management Institute (GMI) and of the GMI Vision. Because I participated in the development of the components that became the core of GMI and because I participated in the development of the GMI concept itself, this work is somewhat a narrative of that development process.

There are several threads that are woven through this narrative. There are at least five innovative key products or services that were developed by the GIS Management Institute, each with its own origin, concept, development, review, and implementation.

The GMI Vision itself evolved over time. There was a major, parallel outreach and communication program, targeted to the international GIS professional community, which is also a part of the narrative of GMI development.

There was continuous and close coordination with the URISA Board in the development of GMI. At several points the URISA Board took a variety of actions related to the development, scope, vision, and future of GMI. To aid in understanding its development, evolution, and demise, a chronology has been provided to document key GMI-related activities and milestones (figure 1).

To aid in understanding the development of individual GMI components and of the GIS environment in which they were developed, I have tried to focus on several themes woven throughout this work.

I have tried to describe the challenges that GIS managers continue to struggle with. The theme of professional GIS management is a complex responsibility, with mastery or at least competence in many domains of knowledge required for success. I have tried to describe some of the key challenges faced by GIS managers and to outline how GMI was developed to address them.

Another theme is the progress of academic and practitioner-based research and writing in the areas of GIS operations and management. I have tried to present a survey of this research and writing before and during the time that the GMI was in development.

Another theme relates to outreach to the GIS community to describe the development of GMI components and to invite other GIS professionals to join in their development. Indeed, GMI would not have been developed without outreach to describe the original GIS Capability Maturity Model.

I try to describe the components of the GMI in sufficient detail so that a knowledgeable GIS manager or stakeholder can understand its basic utility. This theme is intended to encourage the reader to explore each component in more detail to determine its usefulness within their individual GIS environment.

Another theme that may seem tedious to the reader is to document the development of the GMI and its components in detail. The development of the GIS Management Institute was the work of several dozen dedicated but otherwise un-recognized GIS professionals. The references and end notes serve to fairly document the many individual GMI volunteers, whose contributions have not otherwise been publicly recognized on a comprehensive basis.

URISA has been at the forefront of research, education, and support of GIS organizations and GIS operation and management since its early formative stage. As geographic information systems and science were first being conceived and promoted more than 50 years ago, questions of the best means to apply GIS for organizations have arisen.

URISA has a long record of leadership within the domain of GIS management (Wellar 2012). URISA established the GIS Management Institute because of the need to support GIS managers, to provide management resources for organizations that operate GIS, and to create a structure for those who do research, teach, and consult in the GIS management field.

The prominence of the GMI has been progressively downgraded by URISA. In 2016 the GIS Management Institute was described on the website of URISA:

The GIS Management Institute (GMI) develops, manages, and implements professional GIS management products, and services, necessary to assess the maturity and capability of GIS operations. It coordinates administrative activities that support, market, and provide GMI products and services as performed by URISA headquarters staff and volunteers.[iii]

In late 2017 GMI was described on the URISA website as:

URISA's GIS Management Institute® (GMI) helps organizations identify and implement enterprise GIS management practice improvements.[iv]

This is a significantly altered and reduced 'purpose and scope' from what was previously proposed by the GIS Management Institute Committee and approved by the URISA Board of Directors. Seven years of significant volunteer effort went in to planning and developing the GIS Management Institute. Every step of the way, the GMI Committee discussed, wrote about, presented, and publicized its work in an open and inclusive professional process. Over seven years, scores of volunteers participated in hundreds of meetings and conference calls, and devoted many thousands of hours of effort to develop the GMI Scope and Vision.

This work will also provide an outline of the level of effort and broad consensus that led to the GMI Vision. Its publication will document and respect the work of those who contributed their expertise and extensive volunteer time and their intellectual endeavor to the development of the GIS Management Institute Vision. Lastly, it will outline what the GMI might pursue, if URISA chose to again support the GMI Vision in the future.

1
THE GMI VISION

Beginning in 2008 I have been directly involved in creating the core components that enabled URISA to initiate the development of the GIS Management Institute (GMI). The culmination of that work resulted in the adoption of the GMI Vison by the URISA Board of Directors as incorporated into the revised URISA Bylaws adopted on March 16, 2013.

> *204.1.2 GIS Management Institute. The GIS Management Institute (GMI) develops and promotes URISA policies and outreach activities related to advocacy for the GIS profession; formulates professional practice guidelines and standards; evaluates the maturity and capability of GIS operations; develops, maintains, and promotes the GIS profession and GIS management bodies of knowledge; accredits GIS organizations and academic programs; discovers, refines, and publicizes best practices; develops publications and workshops related to the GIS profession and the management of GIS organizations; and advocates for effective GIS policies and funding at the local, state, national, and international levels. GMI may also develop new certification programs that are logically within the scope of its operations. Volunteers and contractors affiliated with GMI need not be members of the Association.*[v]

This statement of the GMI Vision adopted for the URISA Bylaws was based on a much more detailed GMI business plan. This plan envisioned the development of many inter-related products and services that would benefit the GIS profession and the agencies and companies that use GIS. How did this vision come about?

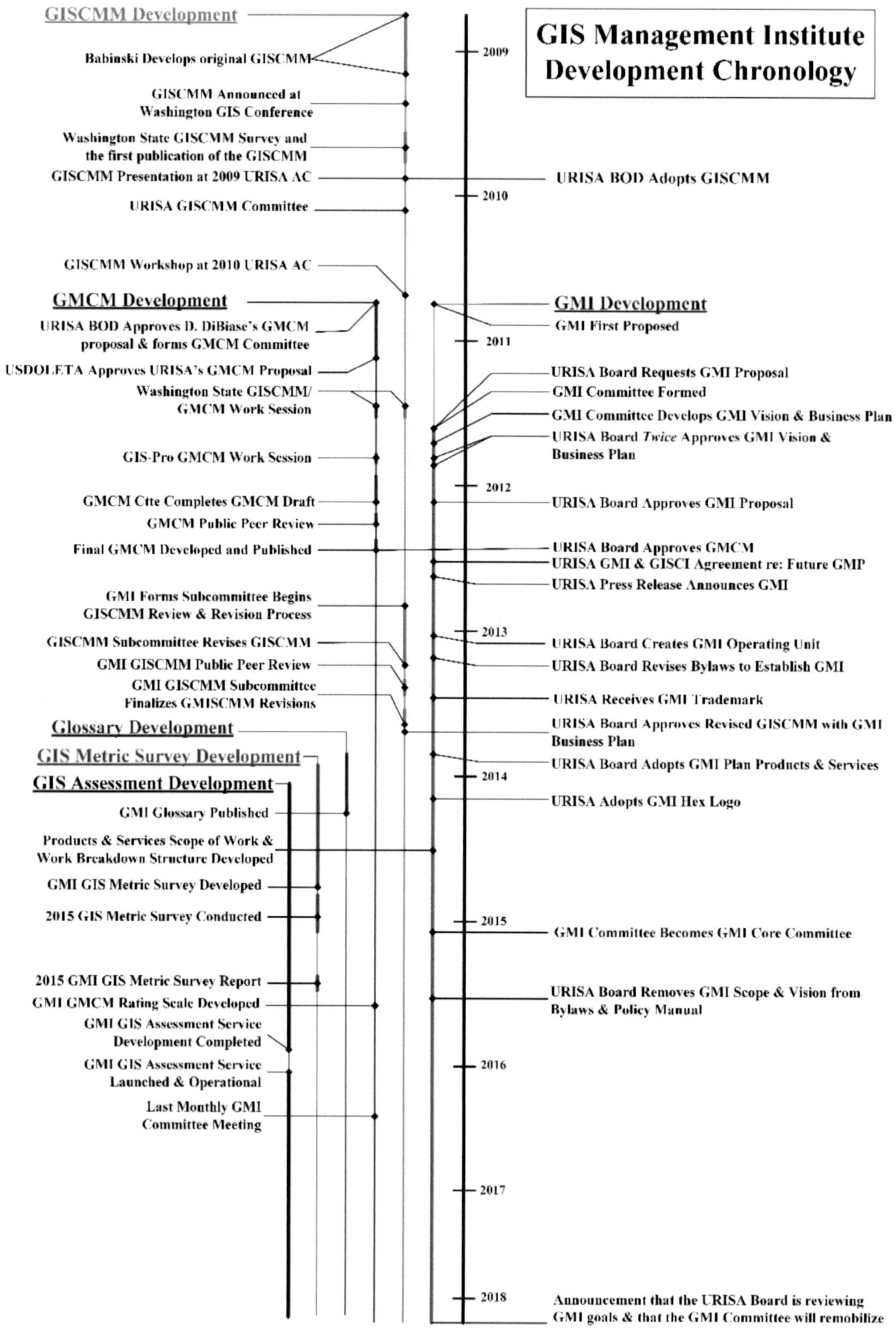

Figure 1
GMI Chronology

2
The Challenges of GIS Management

I began my GIS career in 1989 as supervisor of the GIS Mapping Unit at the East Bay Municipal Utility District (EBMUD) in Oakland, California. An EBMUD colleague and URISA member, Andy Datlen, gave me his collection of URISA Conference Proceedings to use for learning from others who were beginning to conceptualize and put in to practice enterprise geographic information systems for public agencies, utilities, and other organizations. URISA publications, conferences, educational offerings, and professional acquaintances have aided my career as a GIS professional and as a professional GIS manager. I was introduced to the BAAMA Chapter of URISA by Bruce Joffe. Joffe and BAAMA helped me and EBMUD with advice and answers to the problems of GIS development, implementation, management, and operation.

Two significant experiences during my career at EBMUD led to the formulation of questions about GIS operations, management, success factors, and sustainability.

First, I observed that of the 39 cities across parts of two counties within the EBMUD service area, most of their GIS operations struggled and some failed, even after significant effort and investment in the components and initial operation of the GIS. Why were so many organizations not able to successfully develop, implement, and operate sustainable GIS programs? And why was there so much variation in the structure, resources, and use of GIS across similar organizations? I learned that even within this small region, GIS operational and management practices varied significantly.

I joined a small informal group of local public water and wastewater agency GIS managers to share operational problems and discuss our own management 'best-practices'. GIS operational structures and management practices were being developed at that time, largely in isolation, in hundreds of GIS organizations across the world. EBMUD struggled to complete its initial GIS development capital project and subsequent transition to ongoing operations, requiring significantly more time and budget than originally planned.

Second, when EBMUD launched a major project to develop a GIS-based hydraulic modelling computer application, the project manager, Xavier Irias, recommended that EBMUD utilize the newly developed Capability Maturity Model[vi] from the Software Engineering Institute to assess the competency of potential software development consulting firms. The result of this focus was a rigorous, objective vendor selection process and project management approach which led to a successful project and a sustainable implementation.

This success led me to question my observations about GIS management and operations. Could GIS programs apply concepts similar to those of the Software Engineering Institute's Capability Maturity Model to achieve better and more sustainable results?

Origin of the GIS Capability Maturity Model

In 1998 I moved to Seattle, Washington to begin my career with King County GIS. Within my first year, I learned that King County shared some similarities with EBMUD. There are 39 cities within King County, and of those with GIS operations, many struggled and some failed, even after significant effort and investment. I helped form an ad-hoc group of GIS managers from a variety of local public utilities, cities, and nearby counties to share our GIS management challenges and operational success stories.

Soon after I moved to Seattle, a GIS colleague (and former URISA President) startled me by suggesting that: 'GIS is a failure. The investment and level of effort cannot be justified by the benefits that are derived from GIS.'

Despite this dire statement, for the next ten years I worked to help make GIS succeed for King County and to deliver benefits to justify the large investment and level of effort required.

One of my first efforts was to develop a report based on a survey of the departments that were end-users of the King County GIS as it existed in 1998 (Babinski 1998). This report focused on framework data development, including parcel, street network, street address, and digital orthophotography data, as well as other central GIS services. For 22 county agencies it recorded their level of reliance on individual GIS data sets as either legal, integral, critical, or for reference requirements. It identified the number of GIS staff within each agency, as well as the number of GIS end-users. The survey GIS needs rating categories were defined as:

- *Legal = Legal or regulatory compliance need for data/service within department*
- *Critical = Need is critical to the department's work function with no other practical source of data or service*
- *Integral = Need is integral to the department's work, but there may be other sources of data or service that can be pursued*
- *Ref. = Use is for reference purposes only*

I became active in the Washington Chapter of URISA and also began to participate in URISA events at the international level. I began to do some original research related to GIS management practices and GIS operational performance metrics, often using Joffe and URISA Past-President P. Eichelberger for peer-review of my work.

My early research included a GIS knowledge and best-practices sharing proposal (Babinski 2002), GIS organizational structures study (Babinski 2003a), best-practices proposal for GIS marketing (Babinski 2004a), enterprise GIS funding methodologies study (Babinski 2005), innovative practices in GIS management (Babinski 2006), GIS performance metrics (Babinski 2007), and analysis of GIS-based web mapping application usage by internal and external users in cities and counties (Babinski 2008).

A few years after the completion of the county's core GIS capital project, King County GIS was reorganized as a centralized enterprise system. As a King County internal service fund, it was mandated to be financially self-sufficient.

In 2002 the King County GIS Technical Committee commissioned a report on the results of the Capital Project and additional development to date (Babinski 2003b). Typically, a project closeout report is produced at the end of a major project, to document the results of the project in terms of the original scope, schedule, and budget. A closeout report is a standard component of a project's life cycle, as defined by the Project Management Institute.

However, King County's initial GIS development was not accomplished by a single discrete capital project. King County GIS was initially developed over a period of five years from two separate capital projects, in addition to other project funding and operational staff resources. Many agencies first develop their GIS capability via multiple small projects, operational development, resource acquisition, and reorganization. A discrete project-specific closeout report will not provide an adequate assessment of the total GIS capability at a single point in time.

The 2003 King County GIS Report assessed the development of GIS data layers, software, custom applications, infrastructure (hardware and peripherals), and internal staff training. It also assessed the use of components of King County GIS (parcel, street network, and street address data, digital orthoimagery, and other key services), as well as the number of GIS staff and GIS end-users for each county department. Development of this report led me to think systematically about enabling GIS components and the operational characteristics of a successful GIS.

The first Geospatial Technology Competency Model was published in 2001 (Gaudet et.al. 2003). Based on this new model, I conducted research within Washington and Oregon to determine patterns of GIS staffing, mapped to the twelve core roles identified in the 2001 Geospatial Technology Competency Model (Babinski 2004b). Two years later, The Geographic Information Science and Technology Body of Knowledge (GIS&TBOK) was published (DiBiase et.al. 2006).

Both the 2001 Geospatial Technology Competency Model and the GIS&TBOK focused on knowledge areas required for GIS employment, including GIS management roles and competencies. In 2003, the United States Department of Labor Employment Training Administration first identified Geospatial Technology as a distinct employment area and began work on a new Geospatial Technology Competency Model.

In addition to the Software Engineering Institute's pioneer Capability Maturity Model, some work had been done on information technology-focused models. For example, the IT Service Capability Maturity Model (Niessink et.al 2005) focused on helping IT service provider organizations establish their capabilities to aid in informing customers with a clear understanding of their capability. It was also designed to aid IT organizations in developing service process improvement plans.

Pioneer GIS organizational maturity model approaches took diverging directions. One approach looked at organizational implementation stages, from GIS enthusiast, then progressing to department GIS, centralized GIS, integrated GIS, and finally on to enterprise GIS (Mangan 2008). A geospatial enterprise integration maturity model (Sonnen et.al. 2009) looked at an integration-focused model. It identified development phases, from unstructured, to initial, managed, defined, and on to an optimized integration phase. This model defines achievement at each level across a wide range of complex characteristics of a GIS operation within an enterprise.

These early models lacked any useful level of detail related to specific components and resources needed for an effective GIS operation. They also did not take into consideration the effectiveness of the GIS in operation. However, it is deficiencies or gaps in specific components and resources that hinder full GIS capability development. And it is flaws and weaknesses in operational practices that hinder effective GIS operation.

Could a model provide practical insights into the challenges that local government agencies face and help point the way forward to more successful and sustainable GIS programs with specific priority development actions? Could a model identify measurable developmental gaps? Could a model identify measurable flaws or weaknesses in the operational and management practices of an organization? Could a model facilitate peer-agency comparisons and help identify best practices?

How could the profession collectively help GIS managers with programs that struggle, keep at-risk GIS programs alive, and increase the sustainability of GIS operations? The focus of competency models and bodies of knowledge is on the individuals who work in GIS. Could a GIS operation itself be analyzed systematically to identify specific strengths, weaknesses, and risk factors?

The classic conceptual model for GIS implementation is comprised of four elements: data, hardware, software, and staff. Together, these elements comprise the 'system' that provides a variety of automated functionality that can support a body of end-users to meet their mapping and spatial analysis business needs. Could a conceptual model be developed that addressed the complex developmental characteristics of a GIS operation, as well as the multi-faceted processes necessary for optimal GIS performance?

On November 25, 2008, I began development of a systematic framework to model a geographic information system as it operates within and for the agency that it supports (Babinski 2009a). The problem statement and conceptual goal for this effort was defined as:

> *GIS development typically starts as an idea and progresses towards full maturity. However, the reality of municipal GIS operations is that development is limited by available funds. Often GIS starts as a capital project with the system designed to create the 'best GIS possible' with the funds at hand. This development scenario leads to frequent compromise and deferral of many aspects of ideal GIS development in order to 'go operational' quickly and start delivering value for the agency's investment. Even if a GIS implementation project is completed successfully, it does not mean that an agency has a mature GIS, or even a cost-effective GIS operation.*
>
> *GIS professional staff often know that their operation could benefit from enhancement and refinement but funds, staff, or time for further development are very difficult to come by. Enhancements are often developed as part of GIS operations, but rarely on a systematic basis with a desired end state in mind.*
>
> *The purpose of this model is to provide a means for any municipal GIS operation to gauge its maturity against a variety of standards and/or measures, including:*
>
> - *A theoretical ideal end state of GIS organizational development*
> - *The maturity level of other peer GIS organizations, either individually or in aggregate*
> - *The maturity level of the subject organization over time*
> - *The maturity level of the organization against an agreed target state (perhaps set by organizational policy, budget limitations, etc.)*

The innovative core of this new model was to recognize that there are two focus areas for modeling and measuring a geographic information system: first, the elements that make up the GIS itself, and, second, the maturity of the key processes employed by the people who manage and operate the GIS.

There are a variety of elements of a GIS that must be purchased, developed, or otherwise acquired. These were first identified as: 1) technology, 2) data, 3) resources,

4) infrastructure, and 5) GIS professional staff. The first draft model designated these as 'Enabling Capabilities'.

These elements must also be maintained on an ongoing basis. They must be integrated into a complex and often changing information technology infrastructure. GIS professional support staff and end users also comprise an important element of the GIS. Similar to hardware, software, and data, they need to be nurtured, maintained, upgraded, and – sure as death and taxes – GIS staff need to be replaced.

But independent from these GIS resource elements is the question of how they are used. Any tool, be it a farm plow, machine tool, or surgical instrument, will deliver differing results depending on the knowledge, skills, and abilities of the operator. And for complex systems of tools, like an auto assembly plant, gourmet restaurant - or GIS - the procedures that the operators apply, in addition to their knowledge, skills, and abilities, will deliver variable results. The first draft model designated these as 'execution ability' including the '...ability of the staff to maximize use of available capability...' and the '...ability to execute relative to [a] normative ideal.'

The draft GIS Capability Maturity Model (GISCMM) developed over the following two months addressed both focus areas.

The first 2008 draft of the model included 17 Enabling Capability components:

- *Framework GIS data*
- *Business data*
- *Architectural design*
- *Technical infrastructure*
- *Replacement plan*
- *Software maintenance*
- *Backup and security*
- *Application portfolio*
- *Portfolio management*
- *Portfolio O&M*
- *Operational staff*
- *GIS governance structure*
- *Link to agency strategic goals*
- *Comprehensive budget*
- *Funding*
- *Long-term financial plan*
- *ROI analysis and reporting*

A means of rating a GIS operation against each of these components was needed. The 2008 draft model stated:

> *Technology & infrastructure capabilities are gauged by progress towards an ideal state that is specific for each individual component. For each capability a score card defines characteristics that assess progress towards maturity. Each characteristic is self scored to indicate:*

- *Fully implemented*
- *In Progress with full resources committed*
- *In progress but with only partial resources committed*
- *Planned and resources assigned*
- *Not planned and/or no resources assigned*
- *Not applicable*

The 2008 draft model also named 15 Execution Ability components:

- *Client service development and evaluation*
- *User support, training, and help desk*
- *Service delivery tracking and oversight*
- *Service quality assurance*
- *Application development methodology (CMM?)*
- *GIS project management methodology*
- *GIS system management*
- *Process event management*
- *Subcontract and supplier management*
- *Regional collaboration*
- *Staff development*
- *Policies, procedures, standards, and best practices*
- *Performance measures*
- *Cartographic standards*
- *Business focus (needs assessment)*

The Execution Ability components were rated against the classic CMM process maturity rating scale:

Level Five: Optimized

Level Four: Managed

Level Three: Defined

Level Two: Repeatable

Level One: Ad-hoc

In February 2009 the draft GISCMM was distributed for peer review to G. Horning, King County GIS Center Manager, G. Hocking, Chair of the King County GIS Oversight Committee, L. Owen, King County GIS Programmer, R. Gelb, King County Department of Natural Resources Performance Manager, and M. Jacobson, King County Executive Office Performance Manager.

Based on this King County peer review, one additional Enabling Capability component was added:

- *Professional GIS management.*

One Execution Ability component was also added:

- *QA/QC procedures.*

The peer review group also suggested the addition of a brief descriptive section for each Enabling Capability and Execution Ability component, to aid in understanding the specific definition of the component.

The revised draft GISCMM was presented at the 2009 Washington GIS Conference in Bellevue, Washington (Babinski 2009b). At this event I also announced a proposed survey of Washington State city and county GIS operations against the GISCMM.

During the summer of 2009 the GISCMM was further revised based on the King County peer review and on comments that I received from some of those who attended my presentation at the 2009 Washington GIS Conference.

The GISCMM was revised and expanded. Two Enabling Capability components were deleted or combined elsewhere:

- *Funding (combined with budget)*
- *ROI analysis and reporting*

Five Enabling Capability components were added for a net increase of three.

- *Framework GIS data maintenance*
- *Business GIS data maintenance*
- *Metadata*
- *Spatial data warehouse*
- *GIS staff training and professional development*

Four Execution Ability components of the revised GISCMM were deleted or combined elsewhere:

- *Policies, procedures, standards, & best practices*
- *Performance measures*
- *Cartographic standards*
- *Business focus (needs assessment)*

Two Execution Ability components were added for a net decrease of two.

- *Performance management*
- *Client satisfaction monitoring and assurance.*

Based on these revisions, the first publication version of the model used for the 2009 Washington State survey included 21 Enabling Capability components of a GIS, including data, software, hardware, and staff. The Execution Ability portion of the model included 14 process components that represent the key activities that occur within a GIS so that it can provide functionality to its end-users. Each component included a new brief description to aid in understanding the scope of the Enabling Capability or Execution Ability to be evaluated.

The 21 Enabling Capability components of the published September 2009 GISCMM include:

1. *Framework GIS data: The agency has access to adequate framework GIS data to meet its business needs. For the GIS CMM, framework data is defined as NSDI framework layers (see: http://www.fgdc.gov/framework/)*
2. *Framework GIS data maintenance: One or more data stewards are defined for each framework GIS data layer and the data is maintained (kept up to date) to meet business needs.*
3. *Business GIS data: The agency has access to adequate business data (non-framework GIS data) to meet its business needs.*
4. *Business GIS data maintenance: One or more data stewards are defined for each business GIS data layer and the data is maintained (kept up to date) to meet business needs.*
5. *Metadata: Metadata is available and maintained for all framework and business data layers. Metadata complies with FGDC standards or, alternately, meets internally defined standards that support the business needs of the agency.*
6. *Spatial data warehouse: A centralized spatial data warehouse is available for data stewards to house production framework and business data and for GIS users to access data for GIS applications.*
7. *Architectural design: An architectural design exists that defines the current state and planned future development of the technical infrastructure. The architectural design guides the investment in GIS technical infrastructure, including servers, data storage, applications, network components, etc.*
8. *Technical infrastructure: The technical infrastructure is in place to maintain and operate the GIS. Technical infrastructure includes hardware (servers, storage, desktops, input and output peripherals), network components, operating system, and GIS software.*
9. *Replacement plan: A plan is in place and implemented to replace technical infrastructure components (hardware, network components, imagery) that have a defined 'end of useful life.'*
10. *GIS software maintenance: GIS software is available to meet the agency's business needs. If commercial software is used, it is under maintenance to ensure long term support and development. If open-source' GIS software is used, an alternate maintenance, support, and development capability is available.*
11. *Data back-up and security: A computer back-up system is in place to ensure the security of GIS data and applications. The backup system is tested periodically by running both automated routines and manual processes to restore sample data. System security is in place to control internal and external access to GIS data and applications as appropriate.*
12. *GIS application portfolio: A portfolio of custom or commercial off-the-shelf GIS applications is available to meet the business needs of GIS clients.*
13. *GIS application portfolio management: The agency's GIS application portfolio is managed to a common design and development framework.*

14. *GIS application portfolio operations and maintenance: The agency's GIS application portfolio is kept viable via ongoing support and application maintenance.*
15. *Professional GIS management: The agency GIS is managed by a dedicated, professional GIS manager.*
16. *Professional GIS operations staff: The agency GIS is operated and maintained by an adequate staff of GIS professionals.*
17. *GIS staff training and professional development: The agency GIS manager and other professional staff have access to on-going training to maintain and develop their technical and operational knowledge, skills, and abilities.*
18. *GIS governance structure: The agency has a formal GIS governance structure that links the GIS operation both to users (typically via a GIS technical committee or user committee) and to key decision makers (typically via an oversight committee).*
19. *GIS is linked to agency strategic goals: The GIS exists as a defined organizational unit of the agency with a clearly defined role in supporting the strategic goals of the agency.*
20. *GIS budget and funding: The GIS unit develops a comprehensive budget and has dedicated funding for its operations, including (at a minimum) labor, hardware, software, data, consulting, and training costs.*
21. *GIS financial plan: The GIS unit has a financial plan that includes a funding model (where the money is coming from) and that also projects future episodic costs for software, equipment, consulting, imagery replacement, etc.*

The 21 Enabling Capability components were defined as assets, resources, or infrastructure.

The 14 Execution Ability process components of the published September 2009 GISCMM include:

1. *Client services evaluation and development: How does the GIS unit evaluate agency business needs for GIS services and develop new client service requests?*
2. *Service delivery, tracking and oversight: How does the GIS unit track, monitor, and evaluate client service delivery?*
3. *Service quality assurance: How does the GIS unit ensure the quality of services provided to clients?*
4. *Application development methodology: How does the GIS unit develop custom GIS applications?*
5. *Project management methodology: How does the GIS unit manage projects for which it is responsible? Projects could be either executed in-house or by an outside contractor.*
6. *Quality assurance and quality control: How does the GIS unit assure a reasonable and appropriate level of quality for projects and for ongoing GIS system operation? System operations include database maintenance and spatial data warehouse processes.*
7. *GIS system management: How does the GIS unit manage the core GIS systems that it is responsible for? GIS system management includes system*

administration, database administration, network administration, system security, data backup, security, and restore processes, etc.

8. *Process event management: How does the GIS unit manage GIS system process events? Typical process events include planned hardware and software upgrades, unplanned hardware failure and data loss and restore events.*
9. *User support, help desk, and end-user training: How does the GIS unit support end users, including user guides, help documentation, training, and ad-hoc help-desk and/or on-site support?*
10. *Contract and supplier management: How does the GIS unit manage its purchasing and contracting processes to ensure the best value for the supplies and services that it acquires?*
11. *Regional collaboration: How does the GIS unit manage regional collaboration to ensure that opportunities to share in the development and operation of data, infrastructure, and applications are pursued, and that the agency's GIS is leveraged to benefit other potential local partners?*
12. *Staff development: How does the GIS unit manage the process of developing its staff to ensure that individual staff member skills are developed appropriate to current and emerging technical and business needs? How does the GIS unit ensure that its staff resources meet its operational requirements for individual GIS competencies, including back-up and succession planning?*
13. *Performance management: How does the GIS unit manage performance, including both individual performance and the performance of the GIS unit as a whole?*
14. *Client satisfaction monitoring and assurance: How does the GIS unit monitor, assess, and assure the satisfaction of its clients? Ideally clients should be surveyed to indicate their satisfaction with individual projects or products and with the municipal GIS operation as a whole.*

Each of the 14 Execution Ability components was described as a question to test if and to what extent the agency achieves the defined ability.

The model was designed to allow a GIS operation to assess and measure itself against these 35 individual components. This provided a useful level of detail that would allow GIS managers or other agency administrators to understand the quality and suitability of their geographic information system for defined business purposes.

The Enabling Capability portion of the September 2009 model adopted an assessment scale for each of the 21 components that had been developed for a State of Georgia Geospatial Assessment (Ayan and Ouimet 2011). This scale, as modified and adapted for the 2009 GISCMM, gauged developmental progression from:

0.0: Not planned, no resources

0.25: Planned and resources available

0.50: In progress, but only partial resources available

0.75: In progress and full resources available

1.00: Fully Implemented.

For the 14 Execution Ability components, the model used the classic capability maturity level progression:

Level 1: Ad-hoc

Level 2: Repeatable

Level 3: Documented and Managed

Level 4: Measured

Level 5: Optimized.

Numerical values for both the Enabling Capability and Execution Ability components were not intended to indicate a mathematical value for a given rating. Within the Enabling Capability components, a component rating of 0.75: (In progress and full resources available) is not assumed to have three times the value to the organization of a component rating of 0.25 (Planned and resources available).

Likewise, within the Execution Ability portion of the model, a component rating of Level 4 (Measured) is not assumed to have an intrinsic value to the organization that is twice the value of a component rating of Level 2 (Repeatable).

The numerical values were merely intended to provide a simple means of comparison of the state of development for each component within the organization at a given point in time. There was also no intention to use the individual numerical ratings to develop an overall average numerical rating for the organization.

On September 2, 2009, Version 1 of the GISCMM was published as the Washington State GISCMM survey was launched (Babinski 2009c). The intention was both to have GIS managers assess their own operations, and to solicit further critique of the model itself from the GIS managers who completed the survey. The survey was sent to 25 county and 38 city GIS programs in Washington State. 12 counties and 19 cities submitted completed surveys. Most also submitted comments and suggestions to critique the GISCMM itself.

The results of the 2009 Washington State GISCMM Survey (figure 2) were reported at the 2009 URISA Annual Conference (Babinski 2009d). After the 2009 Annual Conference, at the URISA Board of Director's[vii] meeting on November 4, 2009, URISA proposed to adopt the GIS Capability Maturity Model that I had developed as a URISA initiative (figure 3).

For the next several months a newly formed ad-hoc GIS Capability Maturity Model committee conducted a small number of conference calls to explore possible next steps for the GISCMM as a URISA initiative.

In January 2010 a plan was agreed by the ad-hoc committee to develop a URISA workshop built around the GISCMM. This workshop committee included H. Perkins, T. McDermott, K. Cargo, and me. A URISA GIS Maturity Workshop proposal for the 2010 URISA Annual Conference was submitted the following month. The workshop plan evolved into a three-hour session to review the GISCMM within the context of GIS management practices, examine some current GIS metrics techniques for municipalities and states, and provide an in-depth description of the GISCMM. D. Ayan joined the workshop committee to contribute her work on the draft State Geospatial Maturity Assessment.

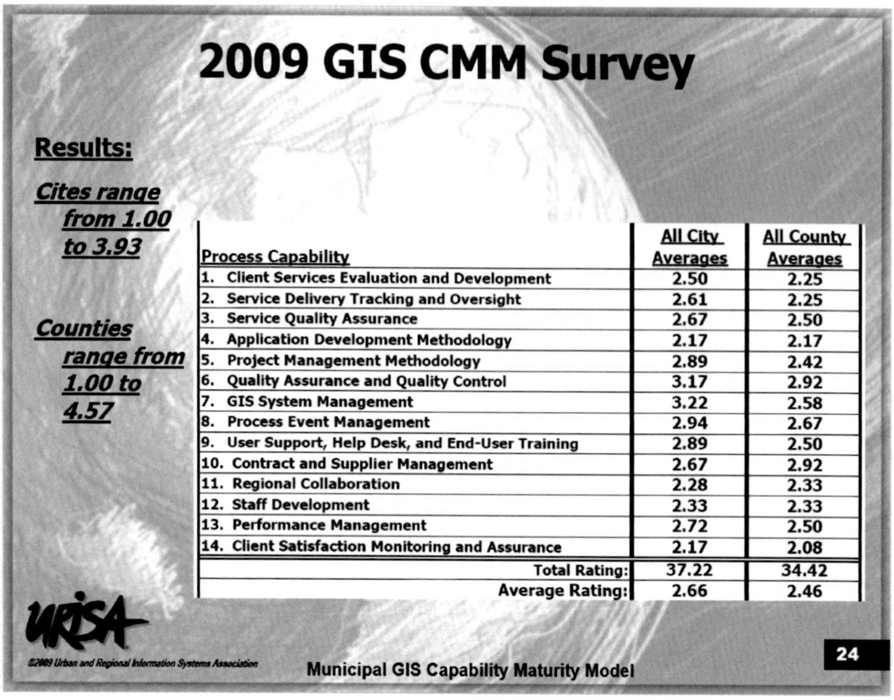

Process Capability	All City Averages	All County Averages
1. Client Services Evaluation and Development	2.50	2.25
2. Service Delivery Tracking and Oversight	2.61	2.25
3. Service Quality Assurance	2.67	2.50
4. Application Development Methodology	2.17	2.17
5. Project Management Methodology	2.89	2.42
6. Quality Assurance and Quality Control	3.17	2.92
7. GIS System Management	3.22	2.58
8. Process Event Management	2.94	2.67
9. User Support, Help Desk, and End-User Training	2.89	2.50
10. Contract and Supplier Management	2.67	2.92
11. Regional Collaboration	2.28	2.33
12. Staff Development	2.33	2.33
13. Performance Management	2.72	2.50
14. Client Satisfaction Monitoring and Assurance	2.17	2.08
Total Rating:	37.22	34.42
Average Rating:	2.66	2.46

Figure 2:
Sample Results from 2009 GIS Capability Maturity Model Survey

The development of the GIS Capability Maturity Model coincided with literature that showed a correlation between business effectiveness and the use of structured management practices. A 2013 study (Bloom et.al. 2013) found that:

'...structured management practices for performance monitoring, targets and incentives are tightly linked to better performance...' but that *'...there is enormous dispersion of management practices across America: 18% of establishments adopt at least 75% of structured management practices for performance monitoring, targets and incentives; while 27% of establishments adopt less than 50% of these practices.'*

Thus, it was believed that the GIS Capability Maturity Model could deliver benefits for GIS, by focusing GIS management on structured management practices and on performance metrics.

GISCMM Outreach 2010

As 2010 progressed, the GISCMM and State of Washington GIS Maturity Survey was presented to various professional groups, including the 2010 GIS in Action Conference in Portland (Babinski 2010a), Washington GIS Conference (Babinski 2010b), and Taiwan Geographic Information Society in Taipei, Taiwan (Babinski 2010c).

Urban and Regional Information Systems Association URISA

URISA BOARD OF DIRECTORS MEETING MINUTES

Date and time of meeting:	Nov. 4, 2009 (11:00 am EST)
Meeting venue:	Via conference call
Minutes recorded and drafted by:	Twyla McDermott, Nov. 5, 2009
Minutes approved by URISA Board:	TBD
5. GIS Capability (Maturity) Model *Greg Babinski*	Could be a potential revenue generator for URISA. Model needs some revision but is being tested at some agencies in the State of Washington. Greg has limited time to evolve this initiative. An ad-hoc committee was formed to conceptualize next steps: Hillary, David, Twyla and Greg. *Action Item: Hillary will convene the ad-hoc committee*

Figure 3
URISA Adopts Babinski's GIS Capability Maturity Model

The National Geospatial Advisory Committee (NGAC) requested a briefing on the GISCMM at its 2010 mid-summer meeting in Washington, DC (Babinski 2010d). NGAC member Jack Dangermond encouraged me to continue developing the GISCMM and invited me to author an article for Esri about its development. This article about the GISCMM appeared in the Winter 2010/2011 *ArcNews*, URISA GIS Management Column, titled: 'URISA Proposes a Local Government GIS Capability Maturity Model' (Babinski 2010e). This article first proposed a theory of GIS management that as the process maturity of a GIS organization increases, its return on investment and associated level of effectiveness increases (figure 4).

The GISCMM was a topic in a President's Column in URISA's *GIS Professional* for the first time (Smith 2010), noting:

> *URISA's work on the GIS Capability Maturity Assessment will enable local governments and other organizations to measure their progress toward geospatial maturity and benchmark that progress against their neighbors and similar organizations around the world.'*

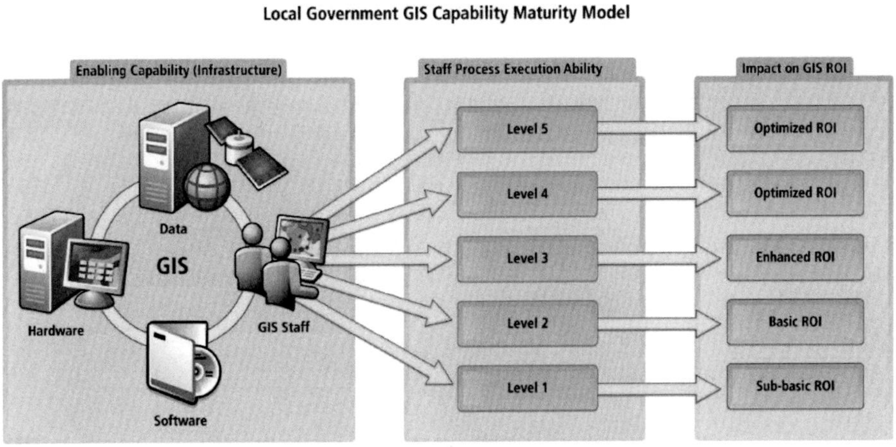

Figure 4
Theory of GIS Management

Origin of the Geospatial Management Competency Model

In July 2010, the United States Department of Labor Employment Training Administration (USDOLETA) had published the Geospatial Technology Competency Model.[viii] At the September 27, 2010 URISA Board of Directors Meeting, Board member D. DiBiase, who had worked on the GTCM (DiBiase et.al. 2010), pointed out that the model had a missing component - Tier 9: Geospatial Management Competency. DiBiase proposed that URISA take on the task of developing the Tier 9 Geospatial Management Competency Model. The URISA Board of Directors approved this proposal.

On September 29, 2010 the GIS Maturity Workshop was conducted in Orlando at the 2010 URISA Annual Conference (Babinski et.al. 2010). After the conference, at the October 1, 2010 URISA Board meeting, a proposal was made to combine the development of the Tier 9 Geospatial Management Competency Model (GMCM) with a comprehensive review and revision to the GISCMM. DiBiase and I agreed to serve as co-chairs of the GMCM Committee (later joined by P. Kennelly). At this meeting, I first proposed that URISA develop a new GIS Management Institute to use both a revised GISCMM and the future GMCM as the basis for the accreditation of municipal GIS programs.

By February 2011, USDOLETA had accepted URISA's proposal to develop the GTCM – Tier 9 Geospatial Management Competency Model. It was agreed that a daylong work-session would be convened at GIS-Pro 2011, with participation by a select panel of highly respected GIS managers, to draft the GMCM. It would then be submitted for open peer review, before final revision and publication by URISA and USDOLETA.

Linking the GISCMM and the GMCM

In February 2011 a plan was developed to prepare the groundwork for development of the GMCM at the planned daylong work session at GIS-Pro 2011 later that year. DiBiase suggested that a strawman draft GMCM should be developed to facilitate the work session at GIS-Pro.

To make the work at GIS-Pro as efficient as possible, I proposed to conduct a preliminary day-long work-session of GIS Managers at the 2011 Washington GIS Conference to accomplish two tasks: a) develop the strawman draft for the GIS-Pro 2011 GMCM work session, and b) conduct a systematic and thorough peer review of the GISCMM, to prepare it for revision and further future use by the GIS community. This proposal was approved by the GMCM Committee.

GMI Outreach 2011

The GISCMM concept, the proposed development of the GMCM, and the idea of a GIS Management Institute were publicized and promoted through the end of 2011. This included a presentation at the 2011 North Carolina GIS Conference in Raleigh, North Carolina (Babinski 2011a) and during my keynote address at the 2011 Alaska State Surveying & Mapping Conference, in Anchorage, Alaska (Babinski, 2011b). The GIS Management Institute and the GISCMM were presented during a daylong 'Law and the Geoweb' Panel Discussion that was part of the 2011 Association of American Geographers Conference program in Seattle (Babinski 2011c). This panel was held on the Microsoft Campus in Redmond, Washington on April 11, 2011, along with panel participants from Google, Open Street Map, U.S. Census Bureau, and Community Commons.

On August 10, 2011, a live URISA Connect Webinar titled 'URISA's Local Government GIS Capability Maturity Model' (Babinski 2011d) presented the development and status of the GISCMM to the entire URISA community.

The GISCMM was also a topic in the President's Column in URISA's *GIS Professional* (Smith 2011a), noting the planned GMCM work session at GIS-Pro 2011 and the role of the GISCMM within the GIS Management Institute. Another *GIS Professional* Article (Terry 2011) indicated that GMI would be a new topic at GIS-Pro 2011.

2011 Washington State GIS Managers Work Session

At the 2011 Washington GIS Conference, on May 9, 2011, in Lynnwood, Washington the URISA GIS Capability Maturity Model and Geospatial Management Competency Model Work Session convened. Participants included:

- S. Beimburn, Manager, Seattle (WA) Public Utilities GIS
- D. Burdick, GISP, Manager, City of Bellingham (WA) GIS
- A. Esnard, GISP, Manager, Multnomah County (OR) GIS
- I. Von Essen, Manager, Spokane County (WA) GIS
- T. Griffin, Manager, Thurston County (WA) Geodata Center
- G. Horning, Manager, King County GIS Center (WA)
- Chair: G. Babinski, GISP, Finance & Marketing Manager, King County GIS Center (WA)

During the work session, participants reviewed the URISA GISCMM, the DOLETA GTCM, and the earlier 'Workforce Development Models for Geospatial Technology' (Gaudet et.al. 2001), which includes extensive management-related competency modeling. The proposed framework for developing the GMCM was described and discussed.

As a starting point, the 2011 Washington GIS Conference GIS Managers GISCMM/GMCM Work Session utilized a draft generic outline for a Tier-9 management competency model from the U.S. Department of Labor Employment Training Administration (USDOLETA 2015). This outline was a simple list of 15 proposed generic management competency areas:

1. *Staffing*
2. *Informing*
3. *Delegating*
4. *Networking*
5. *Monitoring work*
6. *Entrepreneurship*
7. *Supporting others*
8. *Motivating and inspiring*
9. *Developing & mentoring*
10. *Strategic planning/ action*

11. *Preparing & evaluating budgets*
12. *Clarifying roles & objectives*
13. *Managing conflict & team building*
14. *Developing an organizational vision*
15. *Monitoring & controlling resources*

At this point in time there was no industry-specific Tier-9 management competency model that had been developed and published by USDOLETA to guide URISA in the development of a geospatial management competency model. The GIS Work Session managers agreed that the GISCMM capability and process maturity components could help identify GIS management competencies within the framework of the 15 USDOLETA competency areas.

The 2011 Washington GIS Managers Work Session refined the generic USDOLETA list, expanding it to 16 competency categories (URISA GMCM Committee 2011). Within each competency category, the participants added from one to eight 'critical GIS work functions'. As the Work Session participants did their work, they also identified and recorded corresponding components of the GISCMM. The 16 GMCM strawman draft competency categories, along with their corresponding 'critical work functions' included:

A. *Staffing:*
   - *Analyze business needs to determine the technical competencies and the extent of geospatial and other staff required by the GIS organization or project.*
   - *Obtain and deploy geospatial staff with the technical competencies required by the organization or project.*
   - *Leverage staff business, IT, and geospatial skills to support GIS functional business needs.*
   - *Manage staff sub-groups and individuals to support GIS operational areas.*
   - *Corresponds to GISCMM EA 12*

B. *Informing:*
   - *Communicate about GIS technology and operations in an audience-appropriate manner.*
   - *Educate potential clients of business focused GIS capability.*
   - *Inform decision makers of GIS capability within vertical areas.*
   - *Inform decision makers of enterprise GIS capability across vertical business boundaries.*
   - *Communicate business needs, program goals, and project status and needs with GIS operational staff.*

C. *Delegating:*

- *Empower staff to make appropriate decisions*
- *Empower staff to develop their competency by making meaningful and challenging assignments*

D. *Networking*

- *Develop relationships with other organizations to support data sharing, project partnerships, best practices and standards development and adherence.*
- *Advocate and promote shared services.*
- *Interface with enterprise business operations.*
- *Nurture networks of mentors, peers, and colleagues.*
- *Corresponds to: GISCMM EA 11*

E. *Monitoring Work:*

- *Develop, refine, and maintain repeatable processes in alignment with the GIS Capability Maturity Model and best practices.*
- *Measure GIS operational effectiveness via objective metrics.*
- *Measure the use of processes.*
- *Apply project management best practices.*
- *Apply change management best practices.*
- *Apply QA/QC best practices.*
- *Monitor client satisfaction with GIS products, projects, and services.*
- *Corresponds to: GISCMM EA 3, GISCMM EA 4, GISCMM EA 5. GISCMM EA 6, GISCMM EA 7, GISCMM EA 8, GISCMM EA 13*

F. *Entrepreneurship:*

- *Market GIS services or products with an outcome oriented, client-focused approach.*
- *Promote GIS services or products to late adopter agencies within the enterprise.*
- *Develop a 'demand-driven' (versus supply-based) focus to GIS capability development.*
- *Develop long-term client relationships.*
- *Develop strategic partnerships.*

G. *Supporting Others*

- *Analyze business processes to determine GIS needs*
- *Develop service level agreements*
- *Promote a customer service orientation within GIS*

- *Envision and promote cross-business operational efficiencies*
- *Respond to client needs, feedback, and satisfaction.*
- *Evaluate client requests and develop geospatial service response plan.*
- *Corresponds to: GISCMM EA 1, GISCMM EA 2, GISCMM EA 9, GISCMM EA 14*

H. *Motivating & Inspiring:*

- *Think creatively and collaboratively to leverage GIS to solve business problems.*
- *Show GIS staff the 'big picture' role of GIS within the enterprise.*
- *Promote and advocate for GIS.*
- *Promote and advocate for GIS staff.*
- *Look for opportunities to recognize achievements of teams and individuals*
- *Support GIS development in specific business areas:*
    - *Standards development*
    - *Use of and contribution to common data structures*
    - *Subject matter experts in outside business areas*

I. *Developing & Mentoring:*

- *Recruit and hire staff with good geospatial technology knowledge, skills, and abilities and demonstrated competency as life-long learners.*
- *Develop and maintain a geospatial competency needs and staff assessment*
- *Develop and maintain a geospatial staff competency backup, development, and staff training plan.*
- *Promote staff problem solving and creative thinking for leveraging geospatial technology.*
- *Encourage GIS(?) staff to consider cross-business functions in data design and development.*
- *Mentor and support GIS users within end-user departments.*
- *Help individual staff develop a geospatial career development plan*
- *Mentor and support interns and entry level staff.*

J. *Strategic Planning/Action:*

- *Assess business needs and organizational capability*
- *Asses industry trends and their potential effects on the organization*
- *Link GIS to enterprise strategy and direction*
- *Identify metrics critical to tracking a strategic plan upon implementation*

K. *Preparing & Evaluating Budgets:*

- *Develop forward thinking resource requirements aligned with both current operations and future development objectives.*
- *Develop a long-term GIS development financial plan.*
- *Build alliances to develop political and financial support for GIS budgets.*
- *Master the agency's budgeting and budget approval process.*
- *Align budget requests with agency goals*

L. *Clarifying Roles & Objectives*

- *Understand geospatial work functions and assign appropriate competent staff*
- *Understand the utility of planning tools such as the work breakdown structure in defining project roles and objectives*
- *Set performance goals, objectives, and measures for GIS staff.*
- *Clarify role identifiers and suitability for geospatial technology roles.*

M. *Managing Conflict & Team Building*

- *Set conflict management/resolution expectations.*
- *Encourage collaborative, professional conflict resolution.*
- *Promote ad-hoc, self directed teams.*
- *Encourage virtual, multi-agency, multi-jurisdictional teams.*

N. *Developing an Organizational Vision*

- *Understand the GIS organization's role to support the enterprise goals and strategic plan*
- *Develop and promote the GIS organization's vision and mission statement.*
- *Execute the organizational vision with an appropriate mix of hybrid/federated/centralized organizational structure.*
- *Develop best practices and standards that support the organizational vision.*

O. *Monitoring & Controlling Resources:*

- *Set organizational and individual staff performance standards*
- *Implement organizational and individual staff performance appraisals and feedback*
- *Implement procurement and supplier management best practices.*
- *Implement geospatial technology asset management best practices.*
- *Understand the use of project tracking tools such as earned value analysis in monitoring and controlling project progress*
- *Corresponds to: GISCMM EA 10*

P. *Monitoring & Assessing Technology*

- *Monitor new GIS and allied technology and assess its suitability for agency business needs.*

The 2011 Washington GIS Work Session participants agreed that the GISCMM and the proposed GMCM are related in that they both assume many of the same competencies. Components of the GISCMM are much broader and correspond to aspects of the GTCM, including industry-wide technical competencies (tier 4), industry-specific technical competencies (tier 5), occupation-specific knowledge (tier 6), occupation-specific technical competencies (tier 7), and occupation-specific requirements (tier 8), in addition to management competencies.

The 2011 Washington GIS Work Session participants also completed extensive revisions to the GISCMM. The Enabling Capability components increased from 21 to 22, while the Execution Ability components declined to 13. The changed components included:

New EC component: Separate 'GIS Budget and Funding' into two separate components: *EC 20 GIS Budget'* and *'EC 21 GIS Funding*

Deleted EA component: *EA 14 Client satisfaction monitoring and assurance.*

The Work Group also found that '....conceptually, these two separate models intersect in that both the GISCMM and GMCM address many of the same competencies. The GMCM addresses the competencies that individual managers need and the GISCMM addresses process competencies that indicate a mature organization.'

The draft reports of the Workgroup were reviewed and revised based on comments from the participants, as well as by DiBiase, Kennelly, and G. Almvig, Skagit County (Washington) GIS Manager. The new consensus strawman draft of the GMCM developed by the 2011 Washington State GIS Managers Work Group was thus available for planning and preparation for the second Work Session planned for the GIS-Pro 2011 Annual Conference in Indianapolis that November. The May 2011 version of the GISCMM (Babinski et.al. 2011) would also be available for further use and future refinement when the time was appropriate.

3
Origin of the GIS Management Institute Concept

The origin, development, and adoption by URISA of the GIS Management Institute concept occurred in parallel with work on the Geospatial Management Competency Model and refinement of the GIS Capability Maturity Model. URISA's October 2009 strategic planning balanced scorecard included an initiative to: 'Develop/promote Geospatial Maturity Model for member use.' This was proposed to support URISA's strategic objective to 'Promote Geospatial Technology' and '...enhance [its] unique value and relevance'.[ix]

There had been at least two previous proposals to create a special purpose institute within URISA. An institute has been defined as '...an organizational body created for a certain purpose....often they are research organizations created to do research on specific topics.'[x]

In 2009 a proposal was sent to the URISA Board to establish an Institute of Geospatial Professionals (Butler 2009). This proposal was to establish an ANSI-recognized standards setting entity that would focus on professional practice guides for the GIS profession. At about the same time URISA leadership discussed an informal proposal to create a new 'Information Research Institute' which would refocus URISA on both geospatial and non-geospatial information processing applications for government agencies.[xi]

The idea to establish a GIS Management Institute as the home for the GISCMM and possibly the GMCM had been first proposed to URISA in October 2010. This idea grew out of a discussion that I had with Richard Gelb, King County Department of Natural Resources Performance Manager. In 2009, when Gelb first peer-reviewed the original draft GISCMM, he suggested that it could serve as the basis for an *organization* that could 'certify' the capability and maturity of GIS operations. He suggested that the Green Building Council's Leadership in Environmental Engineering Design (LEED) Certification program could serve as a model for an assessing and certifying organization for GIS organizations or operations.

The URISA GIS Capability Maturity Workshop (Babinski, G., Perkins, H., McDermott, T.D., Ayan, D. 2010) presented at the 2010 URISA Annual Conference introduced this concept to the GIS Community. Institutionalization of the GISCMM would:

- *Enhance [the] credibility of results*
- *Promote sound professional practices*
- *Similar to ISO 9000 and CMMI certification*

It was suggested that an agency completing a GISCMM self-assessment could be required to submit a portfolio to validate their ratings, similar to the LEED model.

As early as November 2010, I had proposed to the URISA Board that the Association should focus its publication efforts on the development and publication of GIS management best practices.[xii]

In the summer of 2011, as URISA President-Elect, I chaired the URISA Strategic Planning Committee (SPC). One of the items that I proposed to the SPC was the establishment of the GIS Management Institute as a new URISA program. The SPC approved and forwarded this recommendation to the Board of Directors. At its August 6-7, 2011 meeting, the URISA Board approved a motion to '...create a working group (ad hoc) chaired by Greg Babinski, to undertake his GIS Management Institute (includes GISCMM and Tier 9) initiative and have a status report to the Board by GIS Pro.' At that meeting the URISA Board also asked for '....a vision of what a management institute is and how it ties parts of URISA together...'[xiii]

Founding members of the GMI Committee were C. Smith, D. DiBiase, N. Obermeyer, C. Brown, and G. Babinski, chair. The GMI Committee mobilized quickly and by 8/15/2011 the GMI Committee had a draft 'Proposal to Establish the GIS Management Institute' (Babinski, Brown, Butler, Smith, DiBiase, Obermeyer, and Nelson 2011).

This proposal was discussed with the URISA Board on August 20, 2011. Input and comments from Board members at that meeting included....[xiv]

Bold, pragmatic, haven't heard a better idea for going forward

Aligns with the advocacy agenda

Represents something to shake up the organization

Fits with what we do, but focuses energy

Seems like a next logical step.

Concerns and issues identified at this August 20, 2011 meeting included:

Do we go radical, or morph into a new organization?

How do we align with or compliment what GISCI does?

Does this change our relationship with chapters?

What are the financial implications?

At this meeting, I also raised the need to trademark a variety of terms that potentially should be protected by URISA, including:

GIS Management Institute (GMI)

Certified GIS Management Professional (GMP)

GIS Capability Maturity Model (GISCMM)

GIS Management Body of Knowledge (GMBOK)

This draft was circulated for multiple review cycles to all URISA board members, to all GMI Committee members, and also to the recently elected URISA Board members who would begin their new terms at the end of GIS-Pro 2011.

A key question related to the financial implications for URISA. URISA's business model in 2011 relied on a variety of revenue sources, including membership fees, business sponsorships fees, revenue from conferences, workshop license fees, and minor grants and donations. The major URISA expense item was headquarter staff labor costs. Most conferences broke even at best or lost money.

URISA Chapters did not contribute financially to URISA. Chapter members were not required to be URISA members, although many Chapter members assumed incorrectly that they were automatically URISA members. Chapters also organized their own conferences, which in essence competed with URISA. URISA supported

Chapters via a Chapter Relations Committee and by extending URISA liability and officers' insurance coverage to chapters and their officers.

Another major expense item was publication of the *URISA Journal.* URISA Conferences, the *URISA Journal,* and URISA Chapters were subsidized by membership fees and corporate sponsorships.

Could the GIS Management Institute Vision both provide value to the professional GIS management community and also help URISA's long-term financial viability? GMI planning was based on the premise that by generating some revenue from selected GIS Management Institute products and services, these new URISA financial resources would both contribute to URISA's fiscal health and ensure the long-term financial viability of GMI. And a financially viable GMI would benefit the GIS operations and professional GIS managers that it was being planned to support.

Feedback related to these goals and questions was incorporated into progressive GMI proposal versions on September 7, October 19, and October 23, 2011. Each of these iterative GMI Proposal versions was circulated within the entire GMI Committee for continual refinement.

In parallel with the development of the GMI Proposal, the URISA Board was considering its status as an international organization. The URISA Constitution (URISA 1986) states in Article II, Activities and Scope:

> *URISA is an international Association to stimulate, encourage and otherwise provide for the advancement of an interdisciplinary and multi-professional approach to planning, designing and operating urban information systems. The Association shall operate as an objective educational organization without political, social, financial or national bias.*

A report to the Board of Directors on URISA's future as an international organization (Babinski 2011e) outlined the role of GMI to support the Association's international mandate:

> *The proposed GIS Management Institute (GMI) provides an opportunity to develop a new URISA program with a comprehensive international orientation from the very beginning. The current GMI proposal identifies a large potential worldwide market. GMI best practices, standards, body of knowledge, educational offerings, testing protocols, and membership requirements should be developed to support international participation.*
>
> *To ensure that the GMI is perceived from day one as a non-nation specific professional initiative, the GMI workgroup should include international representation for developing its business plan, budget, and operational proposal.*

2011 URISA Board Action in Support of the GMI Vision

The final draft version of the GMI Proposal was submitted with a GMI Committee Activity Report[xv] dated October 23, 2011 to the URISA Board of Directors for its October 31, 2011 face to face meeting in Indianapolis. At that meeting the GMI proposal was outlined. It included the core Proposal, GMI Vision and Mission statements, and key GMI Objectives:

> Proposal:
>
> *Over the next 2 to 5 years the URISA GIS Management Institute (GMI) should be established as a non-profit professional organization under direct URISA control.*

The GMI will provide a structure to facilitate the re-alignment of URISA from its current broad open membership-based focus to add a narrow but much more compelling GIS management related product and service based focus. URISA, through the GMI, will become the organization that professional GIS managers and entities that operate GIS service organizations look to for 'must-have' certification, products, services, and affiliation.

Vision:

URISA and the GMI are recognized as the worldwide authority for GIS management.

Mission:

The GIS Management Institute certifies the maturity and capability of GIS operations and the professional competency of GIS managers; develops, maintains, and promotes the GIS management body of knowledge; provides GIS management-related education; and advocates for effective GIS management related policies and funding at the local, state, national, and international level.

Objectives:

To achieve its vision and accomplish its mission, the GIS Management Institute will:

- ❑ *Maintain the URISA GIS Capability Maturity Model (GISCMM)*
- ❑ *Administer the GISCMM and certify the competency level of GIS operations*
- ❑ *Maintain the URISA/DOLETA Geospatial Management Competency Model (GMCM)*
- ❑ *Administer the GMCM and certify the attainment of GIS Management Professional (GMP) status*
- ❑ *Maintain and administer the URISA GIS ROI methodology*
- ❑ *Develop and maintain the GIS management body of knowledge (GMBOK)*
- ❑ *Develop educational offerings that will supplement the GMBOK and support professional GIS managers and the development of more capable and mature GIS operations*
- ❑ *Certify outside GIS management educational* programs

The 'problem' that the GMI Proposal intended to address was stated as:

Problems

A) GIS Operational Maturity Varies Greatly, and B) GIS Management Competency is Difficult to Assess

Within the community of GIS there are two distinct but inter-related areas that require attention for GIS to achieve maturity, to maximize ROI from the large investment in and continued expenditures for GIS operations, and to optimize the application of urban and regional information systems. These areas require attention for GIS to progress towards achieving the full potential to society for the major investments in infrastructure, data, and staff development that have been made.

While both of these issue areas exist within national and major state and provincial GIS organizations in the U.S., Canada, and other industrialized nations, they are most prevalent at the level of local agencies: cities, counties, regional agencies, local utilities, service districts, etc. They likely also exist within private industries.

Both of these issue areas represent largely unmet needs within our target market. They provide the potential for URISA to focus its existing programs and energy to develop a dominant and unassailable position to ensure its organizational relevance well into to the Twenty-first Century.

Assessment of GIS Operational Maturity

The first problem that requires attention relates to the great variation in the characteristics of GIS operations in local agencies. GIS has become a common component of city and county governments. All large municipalities and many medium and small sized cities and counties have now established GIS capability.

Just as each municipality is different, municipal GIS operations vary greatly. Partly this variation results from the ongoing development of GIS capability within many city and county governments. But how do these agencies know where their GIS development is in relationship to potential capability for similar agencies?

Other agencies consider that their GIS is mature because their implementation project has been completed and they are doing ongoing GIS operations and maintenance. But how do these agencies know if they are lacking basic GIS capability?

Capability maturity models have been used to assess the ability of agencies to develop software successfully.

The URISA GISCMM has the potential to help GIS operations and the entities that fund them determine their location on the evolutionary path of GIS development. The GISCMM also has the potential to assess the process maturity of the professional staff responsible for the local agency GIS. Because the GISCMM postulates that ROI from GIS investment increases as process maturity increases, local agency decision makers should be motivated to have an independent assessment of their agency's capability and maturity.

Many refinements and improvements to the GISCMM have been suggested. However, for the GISCMM to be a useful tool to help GIS operations continue to make developmental progress and enhance their maturity, it will need to be institutionalized.

Several models for such institutionalization exist, including the Green Building Council (http://www.usgbc.org/) LEED program, and the Capability Maturity Model Institute (http://www.sei.cmu.edu/cmmi/).

Assessment of GIS Management Competency

The second problem relates to the management challenge faced by agencies that use GIS. A city, county, or regional agency of 100,000 population that has been using GIS for 10 years likely has a cumulative investment in excess of $15 million, an annual budget of $1 million, a professional staff of 3-5 people and more than 100 heavy GIS users. However, it is very difficult for an agency to *assess the competency of the manager responsible for administering its GIS and to optimize both the ROI and the effectiveness in support of agency business needs.*

GIS practitioners can be certified via the GISCI GISP program and ESRI's new technical certification program. While there are a few GIS related management programs and extensive literature on the subject, there is no certification program for professional GIS managers. There is also no certification program for GIS management related educational programs.

At GIS-Pro in Indianapolis this year, URISA in cooperation with USDOL will be developing a proposed Geospatial Management Competency Model (GMCM) to complete tier 9 of the recently adopted Geospatial Technology Competency Model.

URISA has the potential to further support GIS operations and the entities that fund them by establishing a program to certify the competency of GIS Managers as a GIS Management Professionals (GMP). The GMCM that will be developed beginning at GIS-Pro can become the basis for a program to assess the competency of GIS managers and promote the focused enhancement of GIS management maturity within the profession.

Similar models for such certification of professional management competency include the AICP program of APA (http://www.planning.org/aicp/index.htm), and the PMP program of PMI (http://www.pmi.org/).

At its October 31, 2011 meeting in Indianapolis the URISA Board took the following action:

Board Decision - Motion to task Greg Babinski to amend by-laws to establish a standing Geospatial Management Committee under the Operations Division and the Strategic Planning Committee to set the initial direction. Motion seconded. Approved unanimously by voice vote.[xvi]

This was the last meeting for three of the 12 URISA Board members. Four days later there was another face-to-face URISA Board meeting in Indianapolis that included the three newly elected Board members. At that November 4, 2011 meeting the development of the GMI was again confirmed unanimously by the URISA Board of Directors as one of the four URISA strategic priorities.[xvii]

Within four days, 15 individual members of the URISA Board of Directors twice unanimously approved the GIS Management Institute, its Vision, and Business Plan.

By early 2012, a Committee of 20 individuals was mobilized to begin work to turn the GMI Vision into reality (see figure 9).

4
COMPLETING THE GMCM AND GISCMM

GMI Completes the GMCM

Preparation for finalizing the Geospatial Management Competency Model occurred as URISA was refining and adopting the GIS Management Institute Vision, Mission, and Business Plan, as described in the preceding chapter.

Between the October 31 and November 4, 2011 Board meetings, URISA's GIS-Pro 2011 Conference took place. On November 3, during the Conference, the second Geospatial Management Competency Model Task Force convened for its day long work session. A month before the Task Force members had been sent the Strawman Draft GMCM developed by the 2011 Washington GIS Conference GIS Managers Work Session the previous May, as well as other reference material.

The Task Force work session was led by David DiBiase. The individual attendees were chosen for their extensive GIS management experience and for the variety of their individual professional focus. To provide some international perspective, representatives from Canada and Taiwan participated. The attendees, with some of their qualifications, included:

- T. Conry, GIS Manager, Fairfax County VA; former Senior Systems Analyst, CH2M Hill; former Senior Project Manager, ERM, Inc.; former Senior Project Manager, Roy F. Weston; member, URISA Board of Directors
- P. Croswell, PMP, GISP, ASPRS-MS. Former URISA Board member and Past-president of URISA. President of Croswell-Schulte IT Consultants.
- D. DiBiase, Manager, World Campus GIS Certificate and Masters programs; GISP; CMS-GIS; Past president, GISCI; lead editor, DoL Geospatial Technology Competency Model; lead editor, GIS&T Body of Knowledge, and URISA Board member. (Co-organizer and facilitator)
- D. Haley, Enterprise GIS Services Lead, Alberta Energy Resources Conservation Board, Canada. Former URISA Board member and Past-president of URISA
- P. Kennelly, Co-organizer, Associate Professor of Geography, Long Island University; Associate Professor of Geography, Penn State; Editor-in-chief, Cartographic Perspectives. (Co-organizer)
- T. McDermott, Corporate Strategic Technology Planning Manager, City of Charlotte NC. URISA Board member
- B. Ryan, CP, PLS. URS Corporation
- R. Somers, Founder and president, Somers-St. Clair; URISA Board rep to GISCI; Past-president, Geographic and Land Information Society; Past chair, American Congress on Surveying and Mapping
- B. Stauffer, Vice president, geographIT
- Dr. Chi-Hong Sun, Chairman and General Manager, Taiwan GIS Center. Past director, Global Change Research Center, National Taiwan University. Past executive secretary, Commission on Sustainable Development Research, National Science Council.

- G. Babinski, GISP, King County GIS Finance & Marketing Manager, former EBMUD GIS Supervisor, and URISA Board member. (Co-organizer)

The end product of the day-long Work Session was a collection of participant comments and notes and a basic framework for a revised GMCM. The team DiBiase, Kennelly, and Babinski worked over the next month to compile a new draft Geospatial Management Competency Model, based on the Work Session notes. This version identified four management perspectives as proposed by Dr. Sun (Managing: Self, Collaboratively, Team, and Work), with 17 competency categories. Within the competency categories there were many individual 'Critical Work Functions.'

During the Work Session, the attendees agreed to continue as the GMCM Task Force, to work afterwards to refine the new draft and to achieve consensus agreement on a version that would be submitted for public peer review. DiBiase led this effort, which went through seven additional revisions.

Finally, by February 21, 2012, the Task Force agreed on a new draft model that identified 74 individual competencies, grouped into 18 'Competency Clusters'. The earlier grouping within four management perspectives was eliminated. On February 22, 2012, the Task Force's consensus draft GMCM was published on the URISA website with a call for public comments. URISA conducted a marketing and social media campaign to publicize the request for comments, which closed on March 31, 2012.

By the close of the public peer review period, 100 responses to the call for comments were received and 94 were found to be authentic. John Johnson of the National Geospatial Technology Center was also invited to submit comments.

DiBiase led the review of comments, compiling a matrix, and conducting another series of review cycles with the Task Force members. By June 2012 the Task Force completed the final review, editing, and consensus approval process and the Geospatial Management Competency Model[xviii] was published by URISA on June 8, 2012 (GIS Management Institute 2012). The final model includes 74 geospatial management competencies grouped into 18 competency clusters. The URISA GMCM was also adopted by USDOLETA as the Tier 9 of the Geospatial Technology Competency Model.[xix]

The 18 GMCM Competency Clusters include:

1. *Geospatial Technology Competencies*
2. *Self-Management*
3. *Human Resource Management*
4. *Performance Management*
5. *Legal Affairs and Policy Management*
6. *Communication*
7. *Team Management*
8. *Relationship Management*
9. *Business Development*
10. *Leadership*
11. *Professional Development*
12. *Strategic Planning and Action*

13. *Work Management*
14. *Geospatial Project Management*
15. *Political Skills*
16. *Contract Management*
17. *Financial Management*
18. *Asset Management*

GMI Completes the GISCMM

As development of the Geospatial Management Competency Model was completed, the GMI Committee began planning a comprehensive review and revision of the GIS Capability Maturity Model.

The GMI Committee began its work on the GISCMM by a comprehensive review of various methods for assessing GIS operations. Performance and management audits are common means for assessing operations within government agencies. An early King County management audit had found that '...using the two criteria of outcomes and budget, GIS projects in King County were completed satisfactorily' (Eklund et.al 1997). In general, management audits do not provide useful inputs for refining GIS operations or for improving GIS performance.

Project closeout reports are another means of assessing a GIS development project. Often a project close out report is a contractual requirement for a government agency capital project. Components of a project closeout report include assuring that all the project work has been completed, identifying any deliverables that were not completed, and serving as the formal completion of a project (Aziz 2015). GIS closeout reports can also be used to help transition a capital development project to ongoing operations and maintenance, review project successes and failures, discuss improvements for future projects, and formalize the end of the project with the customer (Almvig and Burdick 2014).

King County GIS was developed via a variety of individual capital and operational projects at a time when the county was going through a period of merger and organizational transition. In 2003 a special report was commissioned by the King County GIS Technical Committee to determine what was actually achieved. The goal was to go much further than both the 1997 King County Auditor's report (Eklund et.al. 1997) and the May 1998 King County GIS Assessment (Babinski 1998) and focus on actual resources and capabilities that were achieved. Further complicating matters, the original King County GIS Strategic Plan had been scoped as a $22 million project that would deliver a comprehensive GIS database and 126 defined GIS-based applications. However, less than half the required funding was provided.

The 2003 King County GIS report found that there were serious deficiencies in data provided and only about 15% of the 126 planned GIS-based applications were delivered (Babinski 2003b). This validated the need for a standardized GIS assessment mechanism that could be applied at any point in the GIS operational life-cycle.

A Parks Canada performance audit found that Parks Canada '...GIS activities are aligned with and support the Agency's mandate; an appropriate GIS Governance structure exists; and allocated funds are used with efficiency, effectiveness and economy.' (Office of Internal Audit and Evaluation 2012).

There are several other recognized means to assess GIS organizations and/or the organizations that operate a GIS. Business model assessment, organizational assessment, strategic planning, business needs assessment, system data analysis, cost-benefit analysis, maturity modeling, and situation analysis are examples of other GIS assessment types (Somers 2016a).

The GIS Capability Maturity Model uniquely focusses on the GIS resources within an agency and on the performance of the staff responsible for GIS operations. The foundation of the business case for the GISCMM is the theory that more capable and more mature GIS operations can deliver better return on investment and broader and more effective GIS-based business solutions (GIS Management Institute 2013a):

The GIS Capability Maturity Model assumes that mature agencies have more well developed enabling technology and resources, and that their processes and practices maximize the effectiveness of their GIS infrastructure.

The GMI GISCMM Subcommittee issued a call for volunteers to work on reviewing and revising the GISCMM. Eventually 20 URISA members would participate in this effort, including representatives from Canada and Jamaica (see figure 9).

While this group was mobilizing, I compiled a variety of material into a GMI review version of the GISCMM. This included transcribing the notes, comments, and suggestions from the 2011 Washington GIS Conference GIS Managers Task Force for both the rating scales and for each individual component of the model. A cross reference between the individual competencies of the new URISA GMCM to the appropriate component of the GISCMM was included, because organizational process maturity depends, at least in part, on management competency.

The GISCMM Subcommittee volunteers were all GMI Committee members. At the November 27, 2012 GMI Committee meeting it was agreed that the GISCMM review would be conducted at an accelerated pace. A series of GMI meetings focused on the GISCMM review were conducted on December 18, 2012, and on January 8, January 22, and February 5, 2013. Further review via email continued into March 2013. By April 4, 2013, the GISCMM Subcommittee had completed its internal review and agreed to publish the revised GISCMM for public peer review.

The revised 2013 URISA GISCMM added one new Enabling Capability component to the 22 in the model as developed by the 2011 Washington GIS Manager's Work Group:

- *GIS data coordination.*

In addition, the Enabling Capability rating scale was modified to add one additional intermediate rating. A 'not-applicable' rating was also added. This provided users of the model with additional nuance to describe the developmental status of the Enabling Capability components of their GIS Program. The new EC rating scale included:

1.00 - Fully implemented

0.80 - In progress with full resources available to achieve the capability

0.60 - In progress but with only partial resources available to achieve the capability

0.40 - Planned and with resources available to achieve the capability

0.20 - Planned but with no resources available to achieve the capability

0.00 - This is desired, but is not planned

Not Applicable: This is a non-numeric response that requires an explanation of why this component should not be considered in assessing the operation.

The revised 2013 URISA GISCMM Execution Ability components were increased from 13 to 21. The new Execution Ability components included:

EA14. Individual GIS staff performance management

EA15. Client satisfaction monitoring and assurance

EA16. Resource allocation management

EA17. GIS data sharing

EA18. GIS Software license sharing

EA19. GIS data inter-operability

EA20. Legal and policy affairs management

EA21. Balancing minimal privacy with maximum data usage

In addition, EA5 'Application Development Methodology' was changed to 'Application Development or Procurement Methodology' to reflect that many agencies procure some or all of their custom GIS-based business-solution applications.

The GISCMM was posted to the URISA website for public peer review from April to June 2013. There was extensive outreach to the professional community to solicit comments. After the public comment period ended, the GMI GISCMM Subcommittee validated the input received. 28 valid respondents were identified, with 104 individual unique comments and suggestions.

In July and August 2013, the GISCMM Subcommittee considered each individual comment and agreed to various changes. Rationale for action on each comment was documented. One Execution Ability category was added:

EA22. Service to the community and to the profession

The GMI GISCMM Subcommittee also drafted a preamble to the GISCMM that put its intended use into context. This preamble identified that the model would be of use to the GIS community as a stand-alone document for internal self-assessment of an agency or company GIS operation. It was envisioned that GIS staff and managers, as well as agency leadership and stakeholders could use an assessment based on the GISCMM:

To facilitate discussion amongst GIS managers and decision makers who deploy and fund GIS about the appropriate components of a capable enterprise GIS and the characteristics of a well-managed GIS that maximizes effectiveness and ROI from a given level of investment.

The preamble also explicitly put the GISCMM into context within the GMI Vision, Mission, and Business Plan that the URISA Board had twice previously approved on October 31 and November 4, 2011:

The URISA GIS Capability Maturity Model will be used within the GIS Management Institute (GMI) to provide unifying cohesion to other products and services...

The GISCMM preamble identified six key categories of related or dependent GMI products or services:

- *Identify components of the GISCMM that require further research or investigation by the URISA Professional Practices Division. The intended result would be individual GIS management professional practice standards or GIS management best practices.*
- *The compiled collection of GIS management professional practices standards and best practices would form a proposed GMI GIS Management Body of Knowledge (GMBOK).*
- *The GISCMM and GMBOK would also form the basis for a proposed program to assess and accredit GIS operations.*
- *The GISCMM-based GMBOK would include individual best-practice learning objectives, which could form the basis for a GMI accreditation program for institutions with GIS management educational programs.*
- *The GISCMM and GMCM with associated learning objectives and competencies could be used by the GIS Certification Institute (GISCI) to develop a future Professional Geospatial Management Certification program. This program would be developed by URISA but administered by GISCI.*
- *Finally, the GISCMM and GMCM could be used by URISA to strategically focus the development of future workshops, seminars, webinars, and other mechanisms to provide a focused and comprehensive education portfolio to help GIS managers through their careers.*

The GMI Committee reached final consensus approval of the final revisions at its September 4, 2013 meeting and forwarded the final revised GISCMM, with the newly developed preamble, to the URISA Board. At its September 15, 2013 meeting the revised GISCMM, along with the preamble, was accepted as approved by the URISA Board of Directors.[xx] By this action the URISA Board again explicitly reaffirmed its support of the GMI Vision and Business Plan.

The GMI GISCMM Subcommittee members who contributed to the development of the revised GISCMM included:

- G. Babinski, Chair
- J. Albrecht
- C. Anderson
- U. Bera
- C. Brown
- E. Bohard
- K. Brennan
- A. Butler
- D. DiBiase
- P. Eichelberger
- V. Grant
- D. Haley
- K. Mickey

- N. Obermeyer
- H. Perkins
- M. Roche
- C. Smith
- R. Somers
- R. Tomlinson
- W. Nelson, URISA Executive Director

Other non-GISCMM Subcommittee members contributed to the GISCMM review and revision process as well:

- D. Ayan
- A. Esnard
- I. Von Essen
- A. Ibaugh
- G. Terry

The approved URISA GMI GIS Capability Maturity Model was then announced and posted on the URISA website (GIS Management Institute 2013a).

The components of the final approved 2013 GISCMM[xxi] include:

Enabling Capability Components:

1. Framework GIS Data

2. Framework GIS Data Maintenance

3. Business GIS Data

4. Business GIS Data Maintenance

5. GIS Data Coordination

6. Metadata

7. Spatial Data Warehouse

8. Architectural Design

9. Technical Infrastructure

10. Replacement Plan

11. GIS Software Maintenance

12. Data Back-up & Security

13. GIS Application Portfolio

14. GIS Application Portfolio Management

15. GIS Application Portfolio O&M

16. Professional GIS Management

17. Professional GIS Operational Staff

18. GIS Staff Training & Professional Development

19. GIS Governance Structure

20. GIS Linked to Agency Strategic Goals

21. *GIS Budget*
22. *GIS Funding*
23. *GIS Financial Plan*

Execution Ability (Process Maturity) Components:

1. *New Client Services Evaluation and Development*
2. *User Support, Help Desk, and End-User Training*
3. *Service Delivery Tracking and Oversight*
4. *Service Quality Assurance*
5. *Application Development or Procurement Methodology*
6. *Project Management Methodology*
7. *Quality Assurance and Quality Control*
8. *GIS System Management*
9. *Process Event Management*
10. *Contract and Supplier Management*
11. *Regional Collaboration*
12. *Staff Development*
13. *Operation Performance Management*
14. *Individual GIS Staff Performance Management*
15. *Client Satisfaction Monitoring and Assurance*
16. *Resource Allocation Management*
17. *GIS data sharing*
18. *GIS Software License Sharing*
19. *GIS data inter-operability*
20. *Legal and policy affairs management*
21. *Balancing minimal privacy with maximum data usage*
22. *Service to the community and to the profession*

The GMI GIS Glossary

In 2013, as work on the revised GISCMM was progressing, it became apparent to GMI Committee members that there was a critical need for a common understanding of terms used in the GMCM and GISCMM. A GMI Glossary was proposed to define terms as used within the two published documents.

GMI Committee members J. Albrecht and N. Obermeyer volunteered to co-lead a GMI Glossary Subcommittee to develop the 'Glossary of Terms used in the GIS Capability Maturity Model'. Albrecht and Obermeyer were assisted by five other GMI Committee volunteers.

By January 2013 the GMI Glossary Subcommittee had compiled a list of 256 terms used in the two models. The GMI GIS Glossary was not intended to be a complete and authoritative documentation of terms for the GIS profession.

The purpose for the Glossary was intended to assist users of the GMI models:

...interpret the terminology used on a comprehensive basis. The Glossary makes reference to places in the GISCMM and GMCM where the term is cited.' By referring to and applying definitions in the Glossary, individuals and organizations that complete GISCMM and/or GMCM assessments for the GIS Management Institute® would assist in consistent understanding, reporting, rating, analysis, and comparison of assessments.

For each of the terms included, the Glossary cites the location in the GIS Capability Maturity Model or in the Geospatial Management Competency Model where the term is found. Each term includes at least one definition, along with the source of the definition. Some terms have as many as three definitions cited. 17% of the definitions were authored by GMI Committee members themselves. A few were based on defined usage within the GISCMM and GMCM.

As with other GMI documents, an internal peer-review was conducted once the draft was completed. The final approved Glossary was published on May 8, 2014 on the URISA GMI website (GIS Management Institute 2014a).

The GMI GIS Glossary Subcommittee members who contributed to the development of the GMI GIS Glossary included:

- J. Albrecht, Co-Chair
- N. Obermeyer, Co-Chair
- R. Somers
- H. Perkins
- D. Haley
- K. Mickey
- G. Babinski

5
URISA ADOPTS THE GMI VISION

In parallel with the development of the GMCM, the revised GISCMM, and the GMI GIS Glossary, the refinement of the GMI Vision and structure within URISA and the GMI Committee continued.

At its February 24, 2012 face to face meeting in Chicago, the URISA Board considered the GMI proposal. A motion to accept the document as revised as official URISA policy was approved unanimously by voice vote.[xxii]

GMI Outreach early 2012

The development of the GMI Vision continued to be informed by various presentations and publications to the international GIS community. These included:

- On March 14, 2012, the GMI was described at the 2012 joint Oregon URISA and PLSO (Public Land Surveyors of Oregon) GIS-in-Action Conference in Portland, Oregon. (Babinski 2012a)
- In March 2012, a first-of-its kind retrospective GIS return on investment study was completed by a research team from the University of Washington, Evans School of Public Policy, Cost-Benefit Analysis Center. This study, with joint funding by the State of Oregon and King County, Washington, found that over a period of 18 years, King County GIS had delivered at least $776 million in net financial benefits (Zerbe et.al. 2016). This led to interest within the GMI Committee to add a proposed GIS return on investment component to the GMI products and services portfolio.
- On May 3, 2012, the keynote address at the inaugural event of the URISA-UAE chapter in Dubai, United Arab Emirates, titled 'URISA and the Development of a Geospatial Society' focused on GMI development (Babinski 2012b).
- On May 8, 2012 a paper was presented at the 2012 Washington GIS Conference in Tacoma, Washington titled 'URISA Develops the Geospatial Management Competency Model (GMCM) for USDOLETA' (Babinski 2012c).

GMI and GISCI

At its October 31 and November 4, 2011 meetings the URISA Board of Directors had twice unanimously endorsed the development of the GIS Management Institute as outlined in the October 23, 2011 Proposal. However, in early 2012, a flaw was identified with one aspect of the original GMI proposal: the goal of developing a program based on the GMCM to 'certify' professional GIS managers (GMP).

URISA had participated in the development of the GIS Professional Certification (GISP) program, beginning as early as 1993 (Obermeyer 1993). In 2004 the GIS Certification Institute was established to administer the GISP program that had been developed under URISA auspices (Grams 2004). URISA Board member and GMI Committee member A. Butler, who had participated on the URISA GIS Certification Committee, pointed out that URISA has an MOU with GISCI that prevents it from developing competing certification programs. Also, URISA's relationship as one of GISCI's member organizations would create a conflict of interest if URISA pursued a GIS management certification program on its own.

On April 23, 2012, URISA proactively notified GISCI of the possible conflict of interest within the broad scope of the GMI proposal.

On June 8, 2012, GISCI Executive Director S. Wilson responded, stating that "The GISCI Board of Directors believes that URISA accreditation of GIS operations and GIS educational programs is a viable and worthy part of URISA operations and future direction and both will contribute greatly to industry growth. However, we have a problem with URISA certifying GIS professionals in any area of expertise."[xxiii]

M. Van Hook, GISCI President, and I, as URISA President, agreed to discuss a mutually beneficial way forward related to the development of the GMI and the shared goal of a future professional GIS management certification program. On June 18, 2012, Van Hook and I issued the following joint statement to our respective Boards:

As agreed in discussion between the Presidents of URISA and GISCI, the following programs within the URISA GIS Management Institute proposal do not represent a conflict of interest with GISCI:

- *Development of a URISA GIS Management Body of Knowledge*
- *Maintenance of the Geospatial Management Competency Model (GMCM)*
- *Accreditation of the capability and maturity of county, city, and regional GIS operations against the URISA GIS Capability Maturity Model*
- *Accreditation of GIS Management educational programs for alignment with the URISA GIS Management Body of Knowledge*
- *Accreditation of GIS Management consultant businesses (not individual professionals)*

The GMI would likely be established as a distinct operational unit within URISA, possibly as a separate 501(c)(6) entity.

The GMI would include an advisory council, of which GISCI would be invited to provide representatives. A key role of the GMI advisory council would be to align GMI and GISCI programs and activities. Other agreements, between the GMI, and in addition to the URISA MOU, may be necessary to support the direct business need to maintain a manager certification designation such that GMI formally recommends the evaluation criteria to GISCI to incorporate into the designation and application/renewal process.

GISCI would work in partnership with URISA's GMI to establish a GISP Management certification designation (GISP-M) and related accreditation programs. The GISP-M would reference the GMCM and the URISA GIS Management Body of Knowledge. Unlike the present GISP, a GISP-M should be required to maintain membership in a national professional organization like URISA, GMI, or other GISCI Member Organizations. It is recommended that other qualified organizations for GISP-M membership would have a substantial role in defining GIS management best practices and standards and in providing educational programs to teach to the standards and best practices.

-Mike Vanhook, GISP, GISCI President
-Greg Babinski, GISP, URISA President[xxiv]

This agreement formed the basis for URISA and the GIS Management Institute to move forward with its Vision, Mission, and Business Plan, without a cloud of conflict or disagreement with the GIS Certification Institute or the GISP program. The GMI

Business Plan was revised to reflect the changes agreed with GISCI related to a future GIS manager certification program.

In 2012, as URISA Delegate to the Coalition of Geospatial Professional Organizations (COGO), I began discussing the development of the GIS Management Institute at the March 2012 COGO meeting. Late in 2012 a COGO 'GIS Certification' subcommittee was formed under the leadership of W. Hodges, GISCI's new Executive Director, to keep all COGO member organizations aware of various GIS certification and assessment development programs.

The GMI Concept Announced

The refined conceptual focus of the GIS Management Institute comprised several inter-related components (figure 5)[xxv].

The two key GMI documents: the GISCMM and the GMCM, would form the foundation used to develop a new GIS Management Body of Knowledge (GMBOK). Individual components of the GISCMM and GMCM would become the topics for the development of individual, peer-reviewed, GMI Best Practices. The collected GMI Best Practices would comprise the GMBOK.

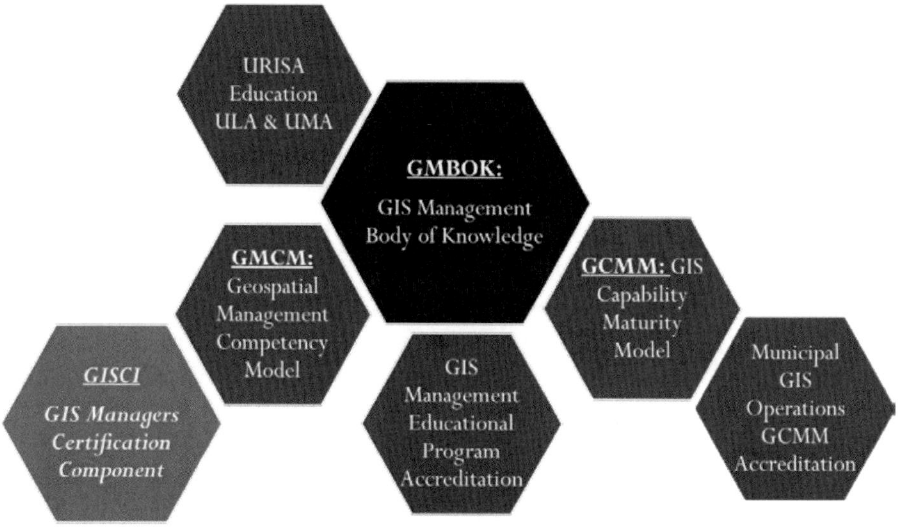

Figure 5
2012 GMI Conceptual Diagram

The GIS Capability Maturity Model would be the basis for development of an assessment and accreditation program for GIS operations.

The GMCM and GIS management-related best practices within the GMBOK would provide the basis for development of a GIS Management Certification program by the GIS Certification Institute.

The GMBOK would serve two additional purposes. First – it would provide focused topics for the URISA Education Committee to develop workshops, publications, webinars, and other resources to address the defined needs of the GIS community. It would also provide topics and best practices that could be incorporated into the URISA GIS Leadership Academy (ULA) and/or form the basis for a possible future GIS Management Academy (GMA).

Second, the GMBOK would provide the professional GIS management competency definition and competency learning objectives that could be used to create the structure and content for a future GIS Management Educational Program Accreditation service.

At the 2012 Esri User Conference in San Diego, on July 24, 2012, the launch of the URISA GIS Management Institute was publicly announced (Babinski 2012d).

URISA Executive Director, W. Nelson and I held a separate URISA/GMI session at the 2012 Esri User Conference. The URISA booth at the conference was staffed to continue discussions with attendees regarding the GIS Management Institute. On July 25, 2012 URISA issued a press release announcing the GIS Management Institute. The press release stated:

> *"The GIS Management Institute (GMI) will develop resources and services that focus on promoting the advancement of professional best practices and standards for the management of GIS operations.*
>
> *"The GIS Management Institute will build upon resources that URISA has already developed, including the GIS Capability Maturity Model, the Geospatial Management Competency Model, the Exemplary Systems in Government (ESIG) Awards, the URISA Leadership Academy, and others. A key component of the GIS Management Institute will be the development of the GIS Management Body of Knowledge (GMBOK).*
>
> *"The GIS Management Body of Knowledge will be the central unifying element of the Institute. It will be used to refine the GIS Capability Maturity Model (GISCMM) and the Geospatial Management Competency Model (GMCM). The GMBOK will be a collection of peer-reviewed best practices and standards that can inform geospatial managers and operations in order to improve the effectiveness of their use of geospatial technology.*
>
> *"The GMI will develop programs based on the GMBOK to accredit the capability and maturity of GIS operations against the GISCMM. It will also develop a program to accredit GIS management educational programs against the GMBOK and GMCM. URISA has agreed to work in cooperation with the GIS Certification Institute (GISCI) to advance the future certification of GIS managers."*[xxvi]

GMI Outreach mid-2012

A *GIS Professional* article (Butler 2012) reflected on the upcoming fiftieth URISA Annual Conference later that year and explained the major reorganization of URISA proposed to help launch the GIS Management Institute with a secure foundation.

Following the public announcement of the launch of the GIS Management Institute, international outreach accelerated:

- On August 6, 2012, a presentation to the Singapore Land Authority GIS Management team outlined the new GIS Management Institute (Babinski 2012e).
- A few days later, on August 13, 2012, the same presentation was made to the leadership team of the Taiwan GIS Center and the Taiwan Geographic Information Society in Taipei, Taiwan (Babinski 2012f).
- That autumn, an article titled 'Geospatial Society, the GIS Profession, and URISA's GIS Management Institute' was published in the URISA *Managing GIS*

Column of *ArcNews* (Babinski 2012g). This article outlined the reason that URISA launched development of the GMI and how it would support professional GIS management and operations.

- GMI reached out to the British Urban and Regional Information Systems Association (BURISA) which published a paper titled 'URISA's 50 Years as an International Organization' (Babinski 2012h). This paper included the status of the GIS Management Institute and URISA's intention for GMI to be an entirely international program.

Major milestones in development of the GIS Management Institute occurred before, during and after GIS-Pro 2012 in Portland, Oregon. On September 30, 2012, I reported on GMI development progress to the URISA Board. At GIS-Pro 2012, J. Dangermond made a substantial financial commitment to URISA specifically to support the further development of the GIS Management Institute.

The connection between GIS maturity and return on investment first proposed in 2011 was explored at GIS-Pro 2012 in a session titled 'Leadership and Management: GIS Maturity and Return on Investment.' Speakers included R. Somers, D. Slavin, J. Calas, and H. Perkins.

On October 2, the GIS Management Institute Committee held its first face to face meeting, attended by 12 of its 20 members and by eight guests.[xxvii] Guests included representatives from the University Consortium for GIS, the Surveying and Spatial Sciences Institute (Australia), and the Executive Director of the GIS Certification Institute. Action items from this meeting included:

Greg will send GMI background information to all attendees. Action items for further development of the GMI include:

a) Update the GMI Proposal to align with the GISCI Agreement, to refine the best practices alignment table, include the 'Hex Diagram' from the GMI PP, etc. (Greg)

b) Contact with Salisbury University (Greg) regarding their interest in aligning with the GMBOK and in GMI accreditation

c) Revise the GIS CMM to align with the GMI and the GMCM (Greg will draft and send to the committee for review)

d) Develop Brief on issues related to the GMI accrediting GIS management educational programs (Jochen, David, and Nancy): Jochen met with Nancy Obermeyer to discuss and feel that the GMI should focus on organizational accreditation and not academic programs. David indicated that there is a small but significant GIS management-focused academic market that would value GMI accreditation. Laxmi indicated that accreditation is in the UCGIS strategic plan and they have formed an accreditation task force. It was agreed that UCGIS and URISA should keep in contact to further explore this area.

e) Refine URISA Labs Concept (Carl, with Clare and Hilary): Carl stated that URISA awards ESIG awards, but with no feedback on success or failures replicating the programs/practices. The GMI will need to know, what is a good or bad practice? The URISA Labs concept is to respond to questions from GMI (or elsewhere within URISA). A 'Lab' will be stood up to perform a literature review and draft a response. Greg said that these draft responses could form the initial versions of GMBOK best practices. It is assumed that

URISA members, chapter members, and URISA Vanguard Cabinet members would be primary Labs recruiting areas. Carl to consider if UCGIS could provide resources for the Labs also.

f) *Registering GMI trademark and logo (Wendy): Wendy indicated that the trademark application has been filed and a logo will be developed.*

g) *Develop GMI Business Plan (Greg): To be developed after proposal is revised*

h) *Develop GMI Operational Plan (Greg): To be developed after proposal is revised*

i) *Develop GMI Financial Plan (Greg): To be developed after proposal is revised*

The GMI Vision

Two days later at GIS-Pro 2012 the revised GMI proposal (URISA GMI Committee 2012) was provided to the Board, including Vision, Mission, and Objectives described as:

Vision:
URISA and the GMI are recognized as the worldwide authority for GIS management.

Mission:
The GIS Management Institute accredits the maturity and capability of GIS operations and, by association, the professional competence of GIS managers; develops, maintains, and promotes the GIS management body of knowledge; accredits and provides GIS management-related education; and advocates for effective GIS management related policies and funding at the local, state, national, and international level.

Objectives:
To achieve its vision and accomplish its mission, the GIS Management Institute will:

- *Maintain the URISA GIS Capability Maturity Model (GISCMM)*
- *Administer the GISCMM and accredit the competence level of GIS operations*
- *Maintain the URISA/DOLETA Geospatial Management Competency Model (GMCM)*
- *Use the GMCM to assess the competence of individual GIS managers as part of organizational accreditation*
- *Maintain and administer the URISA GIS ROI methodology*
- *Develop and maintain the GIS management body of knowledge (GMBOK)*
- *Develop educational offerings that will supplement the GMBOK and support professional GIS managers and the development of more capable and mature GIS operations*
- *Accredit higher education programs that offer specialized degrees, tracks, options, and/or courses in GIS management*

The future Vision for the GIS Management Institute was publicly described in October 2012 in a *GIS Professional* Article titled: 'Reflections on URISA at 75.' In this article (Babinski 2012i), speaking from a hypothetical future URISA 75th Annual Conference, the GMI Vision was described retrospectively as:

> *'...the GMBOK is the core basis for other GMI programs and services. It has been used to update and refine the URISA GIS Capability Maturity Model (GISCMM). It has been used to keep the Geospatial Management Competency Model (GMCM) current. Based on the GMBOK, GISCMM and GMCM, we have accredited hundreds of geospatial management educational programs worldwide.*
>
> *'Based on the GMBOK and GISCMM, the GMI now has a network of thousands of local government agency subscribers to the GMI. GMI subscriber agencies receive GIS Management Institute Accreditation (GMIa) based on their geospatial capability and the process maturity of their geospatial operations. GMI subscriber agencies also receive access to bench marking metrics collected as part of the annual online accreditation process, so each subscriber can compare itself to peer agencies across state/province or around the world, on demand. The GMI and the Geospatial Management Competency Model also enabled development of a professional GIS management certification program.*
>
> *'Within URISA, the GMI led to a new focus for the development of educational offerings. As each GMBOK best practice or standard was developed, the URISA Lab recommended requirements for a workshop to teach to the best practices. This focus allowed URISA to enlist many volunteers to develop new workshops and serve as instructors, delivered either face to face or via URISA Connect Webinars. The webinars in particular were a great innovation which allowed us to enlist workshop authors and presenters internationally. New webinar translation ability further enabled URISA to deliver GMI-focused educational content around the world.'*

GMI Outreach late-2012

Worldwide education and outreach regarding the GIS Management Institute continued:

- On November 14, 2012, at the Sixth URISA Caribbean Conference in Montego Bay, Jamaica, a lecture was delivered on the development of the Geospatial Management Competency Model (GMCM) for USDOLETA by the URISA GIS Management Institute (Babinski 2012j).
- The following month in Beijing, China, I delivered a lecture on December 5, 2012 at Tsinghua University's Center for Earth Sciences, Redbud Forum on Global Change Science. This lecture, titled 'URISA and the Development of a Geospatial Society: Past, Present and Future' (Babinski 2012k) discussed the development of the GMI, with its foundation based on the GIS Capability Maturity Model and the Geospatial Management Competency Model. It outlined the future potential of the GMI to advance the development of GIS professional management worldwide.

6
URISA Reorganizes to Implement the GMI Vision

At the January 3, 2013 meeting of the URISA Board of Directors, a motion passed unanimously to create the GIS Management Institute as an operating division of URISA.[xxviii] At its face-to-face meeting on March 15-16, 2013, the URISA Board unanimously approved revised bylaws which established the GIS Management Institute.[xxix] Specific sections of the bylaws related to GMI included:

> *204 Operating Units*
> *The Association shall be structured as three operating units: URISA International, GIS Management Institute, and GISCorps. Each of the operating units is led by a core committee consisting of persons representing the Association's membership and the various divisions and standing committees that may be assigned to that unit.*
>
> *204.1.2 GIS Management Institute.*
> *The GIS Management Institute (GMI) develops and promotes URISA policies and outreach activities related to advocacy for the GIS profession; formulates professional practice guidelines and standards; evaluates the maturity and capability of GIS operations; develops, maintains, and promotes the GIS profession and GIS management bodies of knowledge; accredits GIS organizations and academic programs; discovers, refines, and publicizes best practices; develops publications and workshops related to the GIS profession and the management of GIS organizations; and advocates for effective GIS policies and funding at the local, state, national, and international levels. GMI may also develop new certification programs that are logically within the scope of its operations. Volunteers and contractors affiliated with GMI need not be members of the Association.*[xxx]

Thus, the process begun by the URISA Board on November 9, 2009 when it adopted the GISCMM as a URISA initiative, resulted three and a half years later in the URISA Board's approval of both the GMI Vision and of a carefully developed and widely vetted companion business plan to implement the GIS Management Institute.

URISA Adopts the GMI Products and Services Concept

By mid-2013 the GMI plan had been refined to add a level of detail that identified the proposed and inter-related products and services. A new conceptual diagram was developed along with a detailed matrixed description of the proposed individual GMI products and services. The conceptual diagram and products and services matrix was distributed for review in the summer of 2013 and then finalized at the 2013 GIS-Pro Conference in Providence, Rhode Island. The 2013 Conceptual Diagram (figure 6) outlined the key planned GMI products (designated P.1 through P.7), services (designated S.1 through S.8), and miscellaneous products (designated M.1 through M.4) (Babinski 2013k):

P1 - GIS Capability Maturity Model (GISCMM):

- *What it is: 23 Capability Components and 22 Process Maturity Components*
- *Target Market: All enterprise GIS operations, managers, consultants, educators & researchers*
- *Development status and maintenance plan: Complete. Needs annual review with update every three years*

P2 – Geospatial Management Competency Model (GMCM):

- *What it is: 73 individual competencies in 18 competency clusters*
- *Target Market: All enterprise GIS operations, managers, consultants, educators & researchers*
- *Development status and maintenance plan: Complete. Needs annual review with update every three years*

P3 – GIS Management Body of Knowledge (GMBOK)

- *What it is: Comprised of individual GIS Management Best Practices. Estimate 25-35 individual Best Practices that correlate with GISCMM process maturity components and GMCM competency clusters. Each individual Best Practice will include:*
  - *Cross-reference to GISCMM & GMCM*
  - *Description of theoretical best practices*
  - *Reference to best known actual practices*
  - *Indication of most common actual industry practices*
  - *Value proposition for best practices*
  - *Recommendations on implementing and documenting the best practices within an enterprise*
  - *Recommended performance metrics for each best practice*
  - *Digital editable best practice template for internal agency implementation*
  - *Recommended curriculum components*
  - *Bibliography*
- *Target Market: All enterprise GIS operations, managers, consultants, educators & researchers*
- *Development status and maintenance plan: Future - to be developed by URISA Labs.*
- *Value Proposition:*
  - *Each GMBOK Best Practice provides insights and guidelines for enterprise GIS operations to implement peer-reviewed best practices to enhance their effectiveness and ROI.*
  - *Each Best Practice assists GIS managers in their professional development*
  - *Assists educational programs to develop GIS curriculum material*
  - *Assists GIS consultants to market credible management practices*

P4: Develop, maintain and publish the Enterprise GIS Management Accreditation Guidelines (EGMAG). To be developed in the future with S.3 GIS operational accreditation service.

P.5: Develop, maintain and publish the GIS Management Curriculum Accreditation Guidelines (GMCAG). To be developed in the future with S.6 GIS educational program accreditation service.

P.6: Develop and maintain recommended GIS Manager Certification procedures and criteria. To be developed in the future in cooperation with GISCI to allow GISCI to launch a GIS Manager Certification program (S.7).

P.7: Develop and maintain (with others) a URISA GIS Return on Investment Methodology. To be developed in the future with S.8, GIS ROI Assessment Service.

S.1: GMI Website and Accreditation online application.

- *Customer facing custom portal into most GMI services*
- *Supports: S2, S3, S4, & S5.*

S.2: Assessment of enterprise GIS operations development and management.

- *Allows client agencies to compare their enterprise GIS capability and maturity against user-defined peers.*
- *Allows GIS mangers to compare their personal competency against their peers*
- *Allows agencies to compare their metrics against user-defined peer agencies.*
- *Provides access to peer-maintained library of GIS management practices, policies, and procedures.*

S.3: Accreditation of local government agency enterprise GIS based on analysis of capability and maturity against the GISCMM & EGMAG.

- *Allows client agencies to assess the effectiveness of their investment in and operating budget for GIS*
- *Allows GIS mangers assess their development in relationship to an ideal end-state for enterprise GIS*
- *Provides entities and GIS managers with best practice templates for enhancing the effectiveness and ROI of their operations.*
- *Provides GMBOK-based and peer-reviewed accreditation of agency GIS capability and maturity*

S.4: Analysis of local government agency enterprise GIS with development recommendations, based on GISCMM database and benchmark metrics (future).

- *The development recommendations package offered every two years to subscribers provides them with a third-party enterprise GIS development/ enhancement plan.*
- *Low-cost alternative to consultant based recommendations.*
- *Tied to GMI database metrics.*
- *A key task will be development of an algorithm to develop the recommendations package.*

- *Would also likely include a brief executive summary narrative.*

S.5: Access to GMI local agency database for custom analysis (future)

- *Allows ad-hoc analysis of the GMI enterprise GIS database for peer agency comparisons*

S.6: Accreditation of GIS management educational programs based on analysis of program against the GMCAG (P.5). Future.

S.7: Operate (with GISCI) all the components of a GIS Management Certification Program (GISCI confers such certification).

S.8: Operate (with others) a URISA GIS Return on Investment Analysis Service.

M.1: Concise educational requirements and learning objectives: Each GMBOK Best Practice or Professional Standard will require a concise set of learning objectives and educational requirements. These can be used by the GMI to issue an RFP to URISA members to develop the required educational offering. The GMI may also release the RFP to URISA associates, if the required educational offering is not developed within URISA.

M.2: Content for URISA Journal and the URISA Annual Conference: Development of each GMBOK Best Practice or Professional Standard will be presented at GIS-Pro and also documented in an issue of the URISA Journal. These presentations will not provide access to the entire Best Practice or Professional Standard, but will provide sufficient validation of the product to ensure broad awareness and acceptance within the community.

M.3: An extensive database of the characteristics of accredited local agency enterprise GIS operations: This database will be available to GMI subscribers and associates for research and analysis purposes.

M.4: URISA GMI Biennial Conference (2-day event focused on results of GMI programs, products and services. Provides a venue for reporting on research based on the GMI database).

Validation for the concept of a subscription fee-based GIS assessment service (GMI service S.2) and future accreditation services (GMI service S.3 and S.6) came from the Government Finance Officers Association (GFOA)[xxxi] of the United States and Canada and the Chartered Institute of Public Finance & Accountancy (CIPFA)[xxxii] of the United Kingdom.

In 2011 GFOA had announced the launch of a Financial Model (FM) Assessment service based on a model and assessment service previously developed by CIPFA. This proved to be useful almost immediately to help public agencies validate their budgeting, finance, accounting, treasury, and other related functions, against an objective third-party model. Portland, Oregon found it useful to conduct a comprehensive review of financial management, validate competencies, verify performance, and compare against industry best-practices (Bailey et.al 2011).

Roanoke, Virginia used the FM Assessment to highlight educational resources available to improve priority development areas, and to promote collaboration across the agency to chart a path to improved practices (Morrill et.al. 2012).

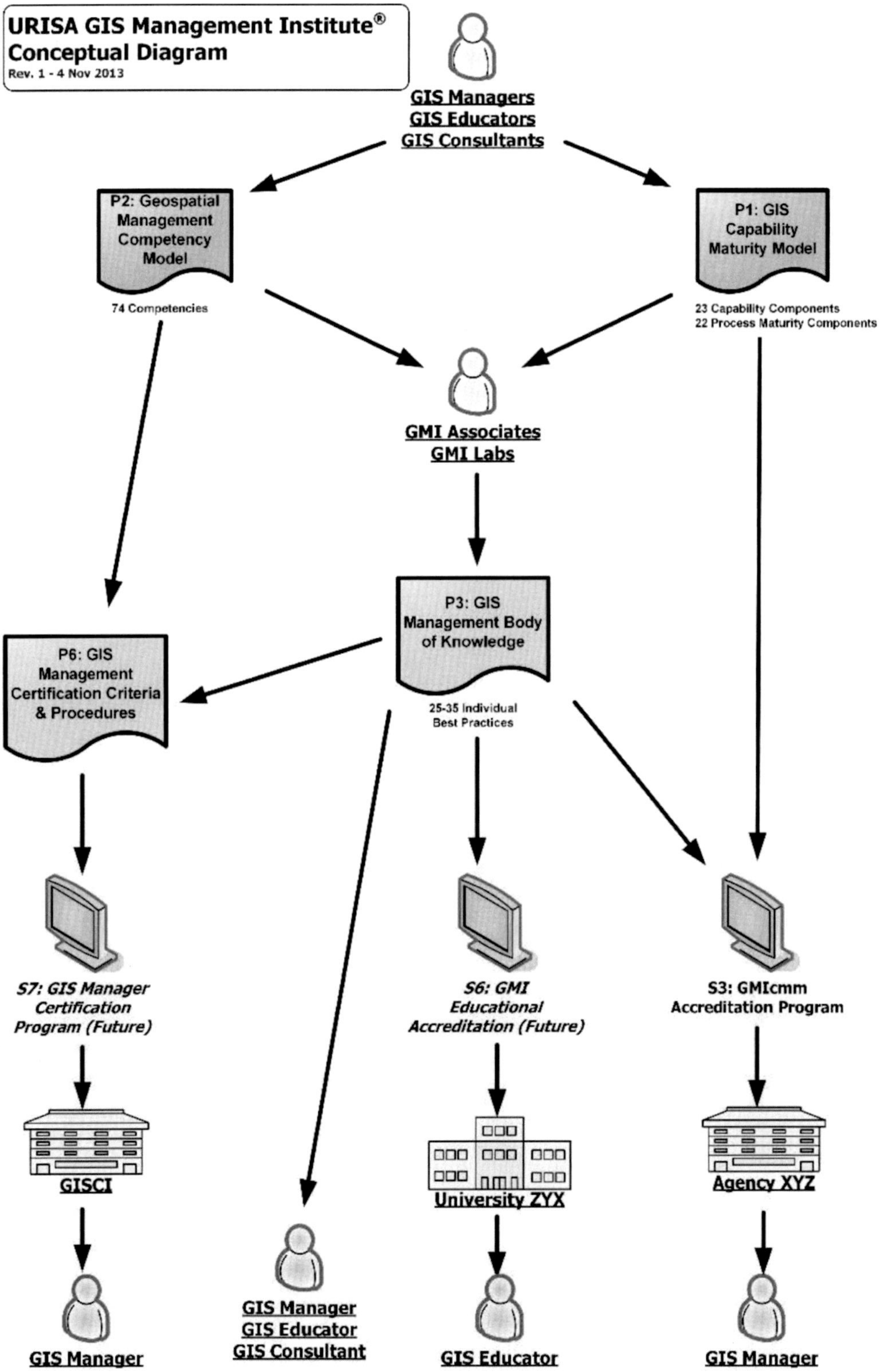

Figure 6
2013 GMI Conceptual Diagram

Further validation of the concept of a subscription fee-based GIS assessment service was provided by the Data Management Association International (DAMA)[xxxiii]. DAMA's mission is:

- *To provide a non-profit, vendor-independent association where data professionals can go for help and assistance*
- *To provide the best practice resources such as the DMBoK and DM Dictionary of Terms in a mechanism that reaches as many DM professionals as possible*[xxxiv]

DAMA also provides an organizational data management professional certification service, based on a Data Management Body of Knowledge (DMBOK).

Future assessment of GIS management-related educational programs within academic institutions (Service S.6) was proposed by D. DiBiase, stating 'GMI's reevaluation of accreditation for organizations that conduct or teach geospatial management is timely and appropriate.' (DiBiase 2012).

As previously described, the GIS Capability Maturity Model as approved by the URISA Board and published in September 2013 also included a preamble that outlined the pivotal role of the GISCMM for the GIS Management Institute. It listed key products and services that were included in the GMI Business Plan. These products and services formed the basis for the reorganization adopted by the Board.

By mid-2013 the GMI Committee had expanded to 23 members (see figure 9).

The GMI Business Case

URISA had adopted the GMI Vision and Business Plan. But for the GMI concept to be viable financially, a business case and detailed financial plan needed to be developed that analyzed URISA current business model with the future impact of GMI. The GMI committee described URISA's current business model as:

URISA has a wide variety of products and service offerings that are recognized within the profession for their quality and value. For five decades URISA has been as an open membership organization, where the interests of individual members (within the very large universe of 'urban and regional information systems') could be discussed, written about, shared, and promoted.

For much of its existence, URISA has been an organization where people who worked with urban information could learn how (if at all) modern technology could be applied to improve government, academic, non-profit, or private sector business processes. URISA's practice has been to call upon its members to write articles and make presentations about their particular work, project, or research. The typical URISA article, workshop, or conference presentation title might be structured within the following formula:

How to apply XXX technology and YYY data for ZZZ business processes.

Because the past 50 years have coincided with the development of cheap and powerful computers and related technology, the potential for individual URISA members (or attendees at URISA events) to learn of some 'game-changing' application of technology that they could apply for their employers or for their careers was great. And because of the novelty and power of technology, how it was applied was often less important than that it was applied – and quickly.

For many of its early years, URISA and its annual conference were the most effective way for an individual to gain exposure to the latest ideas and methods for applying technology to urban information systems. This is no longer the case. Today there are 30 or so URISA chapters with products and services that compete with URISA. GITA and other professional organizations compete head to head with URISA for members and customers. Software companies have their own training, conferences, and publications that provide potential URISA members with an alternate source. Educational institutions, from single classes in community colleges to universities with integrated masters programs also provide alternates to URISA.

Much of the focus of URISA has concentrated on the application of geospatial technology and geographic information systems via conferences, publications, and workshops. URISA played a leadership role in developing the GISP concept and the launching of the GIS Certification Institute. GISCorps is a largely self-sustaining URISA initiative.

Within the last two years URISA adopted the tag line 'The Association for GIS Professionals' and renamed its newsletter 'The GIS Professional.' Both of these actions were taken in recognition that the educational and support needs of current and aspiring GIS professionals align with many of URISA's products and services.

Today URISA would count it as a great achievement to have 2,500 paying members and 500 attendees at its annual conference.

URISA can continue with its existing business model and try to maintain its position as a respected vendor-neutral provider of general products and services that help people learn about and implement information-related technology business solutions.

Or - URISA could build on its existing base but also explore an alternate business model that seeks unmet needs within our target market and focuses its future energy on capturing a dominant and unassailable position related to those needs.[xxxv]

The GMI Proposal was to use the GISCMM, GMCM and other follow-on products and services to provide benefit to local government agencies and organizations that utilize GIS. An analysis of the potential target market for GMI products and services was conducted.

It was assumed that GMI products and services would primarily target medium to large sized jurisdictions. Through the summer of 2013 an analysis was completed of the potential GMI market.[xxxvi]

The GMI Value Proposition was stated as:

GMI products and services will help enterprise GIS operations and those responsible for their development, operation, and management identify and put into place the most cost-effective investments in technology, data, and staff, and identify management practice improvements that will increase agency ROI and maximize effective use of GIS for enterprise business goals.

The GMI Target Markets were described as:

The ultimate primary market for GMI products and services will be any enterprise GIS operation. This is reflective of the fact that GIS provides value to society by combining geospatial technology components with spatial data to allow an enterprise to conduct its business more effectively, develop and provide new services enabled by GIS, while minimizing cost and maximizing net financial benefit.

Secondary markets for GMI products and services will be organizations and individuals who themselves provide support or services to the primary market of enterprise GIS operations. Types of secondary markets will include GIS management professionals, educational institutions that offer GIS management curriculums, GIS management consultants, academics who develop and/or offer courses in GIS management practices, consultants who offer ROI analysis of GIS operations, etc.

A table was also provided with the plan that projected types of organizations and projected total market penetration in one year and three years. Types of organizations targeted included:

- *U.S. cities over 50,000 population:*
  - *Total:* 601
  - *Year 1 target:* 60
  - *Year 3 target:* 150
- *Canadian cities over 50,000 population:*
  - *Total:* 56
  - *Year 1 target:* 6
  - *Year 3 target:* 15
- *Caribbean cities over 50,000 population:*
  - *Spanish speaking:* 54
  - *French speaking:* 21
  - *English speaking:* 22
  - *Dutch speaking:* 2
  - *Year 1 target:* 3
  - *Year 3 target:* 15
- *U.S. counties over 100,000 population:*
  - *Total:* 524
  - *Year 1 target:* 50
  - *Year 3 target:* 125
- *Canadian counties or regional municipalities over 100,000 population:*
  - *Total:* 61
  - *Year 1 target:* 6
  - *Year 3 target:* 15

- *Individual GIS managers:*
    - *URISA members: 221*
    - *GISPs: 992*
    - *Year 1 target: 250*
    - *Year 3 target: 750*
- *GIS management educational programs*
- *GIS management consultants*

The GMI Committee reviewed and confirmed the validity of the target market and the conservative market penetration forecasts outlined. There was agreement that a population of 50,000 was a typical size at which a municipality would have the resources to plan, develop, implement, and operate an enterprise geographic information system. For counties, the critical size was agreed to be a population of 100,000 or more. Of course, some jurisdictions of smaller population might also have the financial resources and enlightened leadership to implement enterprise GIS.

For the GMI Business Plan, a financial plan was developed.[xxxvii] Primary sources of potential GMI revenue included:

- S2: GMI agency GIS assessment fees
- S3: GMI GIS accreditation fees
- P3: Sale of GMBOK
- M4: Biennial GMI Conference
- Contributions and corporate sponsorships
- URISA dues allocation

The revenue was projected to cover a comprehensive set of expenses, including:

- Contribution to URISA Executive Director and Headquarters staff expenses (shared with other URISA operations and programs)
- Contribution to URISA office rent, insurance, and expenses
- Dedicated GMI professional GIS staff
- Website and database costs
- URISA Labs expenses
- Material cost of products
- Travel
- Marketing
- Miscellaneous

Based on conservative market penetration and revenue forecasts, with liberal expense estimates, it was found that GMI would generate a surplus within the first year of operation and provide substantial profit margins in future years.

Later financial analysis was based on these projected target market numbers.

On October 23, 2013, URISA published a GMI Prospectus on the URISA GMI website. The prospectus included the 2013 GMI Conceptual Diagram (figure 6) and described the GMI Program Vision and Mission, Goal, Need, Core Products and Services, and Core Strategy (GIS Management Institute 2013b):

> *Vision and Mission:*
> *The URISA GIS Management Institute® is recognized as the worldwide authority for GIS management best practices.*
>
> *The GMI accredits the maturity and capability of GIS operations and, by association, the professional competence of GIS managers; develops, maintains, and promotes the GIS management body of knowledge; accredits and provides GIS management-related education; and advocates for effective GIS management related policies and funding at the local, state, national, and international level.*
>
> *GMI Goal:*
> *The GIS Management Institute® helps organizations identify and implement enterprise GIS management practice improvements. GIS managers, anywhere in the world, will increase return on investment and maximize the effective use of GIS for their enterprise business goals with GMI products and services.*
>
> *Need:*
> *GIS professionals and practitioners invest considerable time and money for their initial education and continual training, yet recognized professional practice standards and guidelines are lacking in the GIS profession. Public agencies and private entities have invested very large sums of money to develop and operate their enterprise GIS and program specific GIS operations, yet best-practices and investment validation for GIS operations are both lacking. Worldwide, most GIS managers, professionals, and practitioners continue to deliver value to society through the work that they do. But there remains a need for an environment where best practices and professional standards can be developed, validated, and promoted to maximize the value and effectiveness of GIS operations. These are the needs that the GIS Management Institute® will meet.*
>
> *GMI Core Products and Services:*
> *The GIS Management Institute® already has two key products that are central to its core strategy:*
>
> - *The Geospatial Management Competency Model (GMCM) for mangers*
> - *The GIS Capability Maturity Model. (GISCMM) for GIS organizations*
>
> *The GIS Management Body of Knowledge (GMBOK) will be a third key product of the GMI. The GMBOK is intended to be a GMI product that generates substantial revenue.*
>
> *The GISCMM and the GMBOK will also be used to develop an on-line subscription based organizational assessment and accreditation service for enterprise GIS operations anywhere in the world. This will be the primary revenue generating GMI service. Subscribers to the service will populate the GMI database with metrics on their own GIS configuration, maturity assessment, and performance metrics. Their subscription will then provide them access to the GMI database to analyze the effectiveness of individual GIS*

management best practices and to compare their GIS operations against peer agencies worldwide.

The GISCMM, GMCM, and the GMBOK will also be used to develop a revenue-generating, subscription-based GIS educational program accreditation service. The GMCM and the GMBOK will be used in cooperation with GISCI, to develop a revenue generating GIS Management Certification Program.

GMI Core Strategy:

The GMI core strategy is to help those who deploy, operate, and manage GIS organizations enhance their personal competency, and improve the effectiveness and ROI from their investment in GIS. The GMI will mobilize volunteer GIS professionals (to be called GMI Associates) to create the GMBOK, comprised of individual GIS Best Practices. The GMBOK will be developed by starting with frameworks that have already been developed by URISA, such as the GMCM and the GISCMM. Topics for individual GIS Management Best Practices will be developed from the 23 capability and 22 maturity components of the GISCMM. Each topic will include a narrative of the best practice, a policy template, recommended metrics, a description of required professional competencies to support the best practice, and recommended learning objectives to inform the development of a curriculum to teach the best practice.

GMI Outreach 2013-2014

Through 2013 and into 2014 various outreach efforts, both within North America and beyond, promoted the GMI and solicited additional input from the GIS professional community.

GMI Committee members developed articles and delivered a series of lectures about the GMI proposal and development status at a variety of events throughout 2013:

- At the 2013 North Carolina GIS Conference in Raleigh, NC (Babinski 2013a).
- March 11, 2013 keynote address titled 'URISA, The Development of a Geospatial Society and Why GIS Matters' for the Northern Rockies URISA 2013 Intermountain GIS Conference in Boise, Idaho. (Babinski 2013b).
- The following month at the 2013 Surveying and Spatial Sciences Institute Conference in Canberra, Australia from April 15-19, 2013, a lecture titled 'URISA and the Development of a Geospatial Society' (Babinski 2013c).
- On April 29, 2013, at the 2013 Imagin Conference in Mount Pleasant, Michigan, (Babinski 2013d).
- At the 2013 Washington GIS Conference in Lynnwood, Washington, on May 7, 2013 (Babinski 2013e).
- Five days later, at the 2013 Geospatial World Forum in Rotterdam, Netherlands, I delivered two papers related to GMI: Using ROI Methodology to Document the Benefits of King County GIS' introduced the concept of using improved GIS management methodology to enhance GIS ROI (Babinski 2013f). Another paper titled 'The Role of Regional Geospatial Coordination for Effective City Management' on May 13, 2013 touched on the potential role of GMI in advancing more professional GIS management practices (Babinski 2013g).

The GIS Management Institute was a regular topic in URISA's *GIS Professional* newsletter during this period.

- January-February 2013 *GIS Professional* President's Column (Butler 2013b) – prominent discussion of URISA reorganization to pursue the GMI Vision
- March-April 2013 *GIS Professional* President's Column (Butler 2013c) – discussion of how GMI supports the GIS profession.
- May-June 2013 *GIS Professional* President's Column (Butler 2013d) - discussion of GMI development activities; this issue also included a Past-president's (Babinski 2013l) article on interest in GMI from the Survey and Spatial Sciences Institute (SSSI) in Australia.
- July-August 2013 *GIS Professional* President's Column (Butler 2013e) – discussion of URISA's newly agreed organizational structure and its alignment to support GMI operations and vision.
- September-October 2013 *GIS Professional* President's Column (Ibaugh 2013a) – discussion of GMI as a key URISA priority; this issue also included a URISA article (GIS Management Institute 2013c) on publication of the newly adopted, peer-reviewed GISCMM and its role within GMI
- November-December 2013 *GIS Professional* President's Column (Ibaugh 2013b) outlined recent GMI activity and the GMI Vision.

In September 2013, GMI was contacted by the Inter-American Development Bank (IADB)[xxxviii] to discuss the applicability of using the GISCMM for their Emerging Sustainable Cities Initiative. The Bank's interest was to provide a mechanism for gauging the progressive capability and maturity of the GIS operation of medium-sized municipalities across Latin American and the Caribbean that are candidates for financing. A series of conference calls and a webinar were provided for IADB staff that described how the GISCMM could be used to suit their purposes.[xxxix] This activity aligned with the International Proposal approved by the URISA Board in 2011.

After more than 20 years of cooperation between URISA and the Australian Urban and Regional Systems Association (AURISA), a formal MOU was signed in 2010 between URISA and the Surveying and Spatial Sciences Institute (SSSI - the successor to AURISA). Periodic Joint Activity Agreements detailed specific cooperative activities (URISA and SSSI 2012) (URISA and SSSI 2013). For 2012-2014, URISA and SSSI agreed to a variety of GMI-related cooperation.

URISA agreed to provide the following:

- Copies of the current draft(s) of the GISCMM and GMCM
- Selected URISA articles about the GISCMM and the GMCM for inclusion in SSSI publications.
- Copies of conference proceedings and papers related to the GISCMM and GMCM.
- The current URISA President to speak about the GISCMM and GMCM to SSSI members during the URISA visit to SSSI in 2013 and 2014.

SSSI agreed to provide the following:

- Publication of URISA's development of the GISCMM and GMCM in SSSI publications, with request for review and comments by SSSI members to be provided to URISA.
- Participation in a GMCM or GISCMM workshop at URISA's 2012 GIS-Pro conference by the SSSI delegate to URISA.
- A possible daylong GISCMM and GMCM workshop in 2013 and/or 2014 to be conducted by the URISA President during the SSSI visit.

A major article in the *GIS Professional* outlined the proposed reorganization of URISA to accommodate the launch of the GIS Management Institute (Butler 2013a). This article outlined how URISA in-general and GMI in particular would be structured and coordinated to launch, operate, and sustain the GMI Vision.

Another *GIS Professional* article (Anderson 2013) in early 2013 introduced the URISA Labs concept and their planned key role in supporting development of the GMBOK.

Four years of work developing and promoting the GMI concept, Vision, products, and services was reinforced, when on June 11, 2013, The United States Patent and Trademark Office awarded URISA Reg. No. 4,351,832 for the name 'GIS Management Institute,' for Association services, namely, promoting public awareness of best practices and standards in the geographic information systems profession, in Class 35 (U.S. CLS. 100, 101, and 102). In late 2013 a URISA GIS Management Institute logo was developed. The logo design goal was to reflect the GIS Management Institute Vision with its interrelated products and services as depicted in the GMI Conceptual Diagram (figure 5) that I had first developed and used in 2012. The GMI hex-pattern logo (figure 7) was developed by the URISA marketing committee and approved by the GMI Committee in February 2014.[xl]

Figure 7
GIS Management Institute Logo

GMI education and outreach continued through 2013 and into 2014:

- GMI was prominently featured during URISA's 2013 GIS-Pro Conference in Providence, Rhode Island. The President's Program Welcome described the create by the URISA Board of the GIS Management Institute operating unit, with Accreditation, Advocacy, Certification, and Professional Practices Divisions (Butler 2013f).

- At GIS-Pro 2013 in Providence, RI, two papers delivered on September 17, 2013 related to the development and use of GMI: 'Best Practices for Financial Management of GIS Operations' (Babinski 2013h) again showed the potential for utilizing GMI based operational maturity improvement strategies to enhance cost-effective GIS and efficient utilization of financial resources. 'URISA's GIS Management Institute Updates the GIS Capability Maturity Model' (Babinski 2013i) outlined the revised GISCMM and progress in implementing the GMI program.
- A public 90-minute URISA Connect webinar titled 'The GIS Management Institute and the GIS Capability Maturity Model (GISCMM): Updates, Overviews and Implementations' (Babinski 2013j) was recorded live on November 6, 2013 and then made available on the URISA website.
- In January 2014, I traveled to Chelyabinsk, Russia, at the invitation of South Ural State University, to deliver a two-day lecture titled: 'Maximizing Benefits from Municipal GIS Operations The GIS Management Institute and the GIS Capability Maturity Model' (Babinski 2014a).
- On February 14, 2014, I was invited to discuss the GMI and GIS Capability Maturity Model at the URISA Caribbean Chapter High-Level GIS Meeting in St. Marteen.
- In 2014 an Article was published in the *GIS Professional* by A. Ibaugh titled: 'GIS Management Institute: What is it?' (Ibaugh 2014a). Ibaugh stated:

  'The GIS Management Institute® (GMI) is URISA's home for professional practice support to the GIS community. GMI houses three URISA divisions that interactively serve the GIS field through a series of products and services. These divisions will soon offer to URISA membership an on-line organizational self-assessment and URISA assessment program using best practice guidance. One of the key goals of the Institute is to support our membership with evaluating, growing, and improving their enterprise GIS organizations through assessment, performance measures, training, and capacity building.

  'URISA's Professional Practice Division has already begun the process of developing best practice guides for the GIS field and will soon mobilize volunteer GIS professionals (Associates) to begin creating the GMBOK, starting with a topical framework that has already been developed by URISA. Each topic will include a narrative of the underlying best practice, a policy template, recommended metrics, a description of required professional competencies to support the best practice, and recommended learning objectives to inform the development of a curriculum to teach the best practice.'

- January-February 2014 *GIS Professional* President's Column (Ibaugh 2014a) – discussed how GMI and the GISCMM supports excellence in GIS.
- March-April 2014 *GIS Professional* President's Column (Ibaugh 2014b) – prominent discussion about the GIS Management Institute and its role within URISA's future.

- May-June 2014 *GIS Professional* President's Column (Ibaugh 2014c) – mentioned the role of GMI and of the GISCMM within URISA.

7
Development of the GMI GIS Assessment Service

In late 2013 the GIS Management Institute began mobilizing to transform the GISCMM, the GMCM, and the GMI Glossary into a coherent online service for organizational GIS assessment (figure 6, S.3). By 2014 and into 2015, the GMI Committee expanded to 27 members (figure 9).

As planning progressed, several necessary intermediate steps were identified:

- To maximize usefulness, the GMI GIS Assessment Service should allow participants to compare themselves to peer agencies. This meant that the service would require development of a registration mechanism that would categorize agencies.
- By adding a registration form, it was agreed that a moderately detailed survey would be useful, to compile business case information and related metrics about each agency. This would further aid in comparing peer agencies, facilitating periodic update to a subscriber's assessment and comparison with past metrics to show level of progression, and would also be a useful research tool in itself.
- A database and web front-end application would need to be developed to enable users to enter data, run reports, and keep the data secure.
- A legal liability disclaimer and user agreement statement was needed.
- A rating scale was needed for the GMCM.

The development of a GMI GIS database and GMI Subscriber web front-end (S.1 and M.3) was begun by Data Transfer Solution LLC of Orlando[xli] under the guidance of the GMI Committee. A legal liability disclaimer statement was developed by A. Butler and approved by the GMI Committee for incorporation into the GMI subscriber service (Butler 2013g).

The GMI Metrics Survey

A benchmarking study is a common means of gaining insights into differences between peer businesses or government agencies. A benchmarking study is a mechanism to compare a target agency's key metrics, business processes, and performance against similar metrics, business processes, and performance in a limited number of defined peer-agencies.

A benchmarking study is usually conducted by a consultant for a client agency that wishes to gain insights into how its operations compare to its industry peers and to help identify perceived best practices that can be adopted to improve performance.

One early example related to GIS looked at the computer drafting and computer mapping system of the East Bay Municipal Utility District in Oakland, California compared to other major water utilities (Finkle et.al. 1997). Another example was used to compare the City of Charlotte, North Carolina GIS staffing, budgets, hardware, software, database and database management, technology, and software against other local government agencies (City of Charlotte 2009).

A disadvantage of the typical benchmarking study approach is that it is a one-time snapshot with narrow-focus and limited use as time progresses. Bench marking studies are also moderately expensive to perform.

Related to benchmarking studies are comprehensive national or statewide assessments, but they are also expensive and difficult to conduct. An early attempt at a national survey of GIS use in cities and counties (Warnecke et.al. 1998) recognized that knowledge about the use of GIS technology was in its infancy. This study reported on a national survey of a small subset the 83,000 units of local government in the United States. It looked at the incidence of GIS in agencies, the departments within agencies that used GIS, resources deployed, and key applications.

A more recent study (Bradbard et.al. 2017) examined GIS use within a small number of targeted cities and counties in North Carolina and South Carolina. A partial focus of this survey was to identify use of GIS for two broad categories of application: day-to-day operational support and to support decision making and strategic planning.

On November 2, 2013, development began on a concept for a GMI GIS Metric Survey. At the November 19, 2013 GMI Committee meeting, I proposed a detailed metric survey to be incorporated into the GMI GIS Assessment. It would be based on survey data submitted by GIS Assessment Service subscribers. The number of peer-agencies to which a subscriber could compare itself would be limited only by the number of GIS Assessment Service participants. A stand-alone GIS Metric Survey would also represent a new GMI service.

A GMI Metric Survey Subcommittee was formed with N. Obermeyer, C. Anderson, P. Eichelberger, and myself as members. I developed and distributed a draft concept metric survey to this workgroup. By May 2014 the Metric Survey was reviewed, revised, and completed by the GIS Metric Subcommittee and then and approved by the full GMI Committee.

The GMI GIS Metric Survey (GIS Management Institute 2014b) had the following focus and goals:

- *Enable GIS managers and organizations that employ GIS to gain understanding and insights into how GIS are implemented and operated in their organization and other organizations today.*
- *Provide new, comparable, metrics and information from survey results to the wider community of GIS managers, educators, researchers, students, contractors, and consultants to help identify strengths and opportunities for improving cost-effective utilization of geospatial technology within their organizations.*
- *Allow agencies to generate 'peer-agency' analysis – so they could compare their GMI GIS Assessment against other self-defined 'peer-agencies.'*
- *Populate a GMI database that would become a valuable resource to monitor and analyze the progress of GIS development in cities, counties, regional agencies, and other GMI subscribers.*
- *Test the GMI GIS Assessment Service technical infrastructure.*
- *Enable publication of future GMI 'State of GIS' annual reports, similar to the URISA salary survey.*

The components of the survey were organized into nine categories:

1. <u>*Identifying Characteristics (To Classify Subscriber Agency)*</u>
   - *Name of Organization*
   - *Type of Organization*
   - *Location of Organization*
   - *Name and job title of person responsible for GMI Metrics Survey*
   - *Name and job title of person performing internal GMI Metrics Survey validation*
   - *Main Website URL*

2. <u>*Organization Characteristics (To Validate Subscriber Agency Business Use Case)*</u>
   - *Population of Jurisdiction*
   - *Area of Jurisdiction (Square Km)*
   - *Total Property Parcels*
   - *Total Addresses*
   - *Total Utility Customers*
   - *Total Route Miles*
   - *Optional Narrative Description of Organization Business Use Case*

3. <u>*Organization Resources*</u>
   - *Total Number of Employees*
   - *Total Organization Budget*

4. <u>*GIS Organization Resources*</u>
   - *Year Enterprise GIS Development Began*
   - *Total Number of GIS Professional Staff*
   - *GIS Managers*
   - *GIS Programmers*
   - *GIS DBA's*
   - *GIS Analysts*
   - *GIS Technicians*
   - *Other GIS Staff*
   - *Total Number of GIS staff who have any recognized GIS Certification*
   - *Annual GIS Staff Labor Budget*
   - *Annual GIS Hardware Budget*
   - *Annual GIS Software Budget*
   - *Annual GIS Contractors and Consultants Budget*
   - *Annual GIS Data Acquisition Budget (Includes Imagery)*

- *Annual GIS Training and Professional Development Budget*
- *Other GIS Budget*
- *Total GIS Budget*
- *GIS Organizational structure*
- *Organizational Placement of GIS Management and Main GIS Support Unit*
- *Do you participate in regional GIS consortia?*

5. *Data Resources*
   - *Imagery*
   - *Vector Data*
   - *Metadata*

6. *Hardware Resources*
   - *GIS Data storage capacity*
   - *GIS Data storage strategy*
   - *GIS Application Server Strategy*

7. *Software Resources*
   - *Operating system*
   - *Primary GIS software solution*
   - *Secondary GIS software solution*
   - *Database software solution*
   - *Other key software components*
   - *Is all GIS software under maintenance?*

8. *GIS Performance Metrics (Internal)*
   - *Total Number of GIS Users Supported*
   - *Number of GIS Users Supported (Desktop)*
   - *Number of GIS Users Supported (Mobile)*
   - *Number of GIS Users Supported (Via Web Mapping)*
   - *Total annual internal web mapping user sessions*
   - *Total annual internal web mapping hits*
   - *Do you conduct internal customer satisfaction surveys?*
   - *Does GIS provide in-house training to end-users?*
   - *Does GIS provide helpdesk support to end-users?*
   - *Does GIS provide 24x7 GIS system support?*
   - *Annual GIS system availability (%) during business hours*

9. *GIS Performance Metrics (External):*
    - *Do you provide external (public) web mapping application access?*
    - *Total annual external (public) web mapping User sessions*
    - *Total annual external (public) web mapping map hits*
    - *Do you conduct external customer satisfaction surveys?*
    - *Do you provide public access to GIS data?*
    - *Do you charge for public access for GIS data?*

The components of the Metric Survey were approved by the GMI Committee by November 2014 and loaded into the GMI web front end and database design. A free GMI GIS Metric Survey was launched in December 2014. The results of the survey and partial analysis were presented in a paper titled 'The State of GIS in Washington & Oregon: The 2014 GMI Metric Survey' (Babinski 2015a), at the 2015 joint Washington URISA-Oregon URISA GIS in Action Conference in Vancouver, Washington on May 6, 2015.

GMI GIS Assessment Service Technical Development

In 2015 the results of the Metric Survey process and feedback from survey participants were used to refine the GMI GIS Assessment Service technical infrastructure database and subscriber front end. A transcription of the components of the complete GMI GIS Assessment service was developed for loading into the GMI subscriber front end and to format the GMI data base.

The complete GMI GIS Assessment includes five major sections:

1. GIS Metric Survey
2. GISCMM Enabling Capability components
3. GISCMM Enabling Capability Data components
4. GISCMM Execution Ability components
5. GMCM Management competency clusters

The data components of the Enabling Capability portion of the GISCMM were broken out in the survey to allow agencies to provide a rating for each individual framework and business data set. The averages of the data components entered in section 3 are used to populate the four data related components of section 2 (EC1, EC2, EC3, and EC4). All other ratings are entered into the model by the subscriber agency as individual rating numbers. Agencies are not allowed to otherwise enter mid-scale ratings or averages.

The Geospatial Management Competency Model as originally published did not include a rating scale. To facilitate assessing the competency of a GIS manager or of a GIS management team, a rating scale needed to be developed. I proposed adopting a modified version of a Management Competencies Proficiency Scale utilized by the National Institute of Health, Office of Human Resources.[xlii] As modified for the GMCM these competency ratings include (GIS Management Institute 2015):

Not Applicable:

You are not required to apply or demonstrate this competency. This competency is not applicable to your position.

Level One: Fundamental Awareness (basic knowledge):

You have a common knowledge or an understanding of basic techniques and concepts.

- *Focus is on learning*

Level Two: Novice (limited experience):

You have the level of experience gained in a classroom and/or experimental scenarios or as a trainee on-the-job. You are expected to need help when performing this skill.

- *Focus is on developing through on-the-job experience;*
- *You understand and can discuss terminology, concepts, principles, and issues related to this competency;*
- *You utilize the full range of reference and resource materials in this competency.*

Level Three: Intermediate (practical application):

You are able to successfully complete tasks in this competency as requested. Help from an expert may be required from time to time, but you can usually perform the skill independently.

- *Focus is on applying and enhancing knowledge or skill;*
- *You have applied this competency to situations occasionally while needing minimal guidance to perform successfully;*
- *You understand and can discuss the application and implications of changes to processes, policies, and procedures in this area.*

Level Four: Advanced (applied theory):

You can perform the actions associated with this skill without assistance. You are certainly recognized within your immediate organization as "a person to ask" when difficult questions arise regarding this skill.

- *Focus is on broad organizational/professional issues;*
- *You have consistently provided practical/relevant ideas and perspectives on process or practice improvements which may easily be implemented;*
- *You are capable of coaching others in the application of this competency by translating complex nuances relating to this competency into easy to understand terms;*
- *You participate in senior level discussions regarding this competency;*
- *You assist in the development of reference and resource materials in this competency.*

Level Five: Expert (recognized Authority):

You are known as an expert in this area. You can provide guidance, troubleshoot and answer questions related to this area of expertise and the field where the skill is used.

- *Focus is strategic;*

- *You have demonstrated consistent excellence in applying this competency across multiple projects and/or organizations;*
- *You are considered the "go to" person in this area within your organization and/or outside organizations;*
- *You create new applications for and/or lead the development of reference and resource materials for this competency;*
- *You are able to diagram or explain the relevant process elements and issues in relation to organizational issues and trends in sufficient detail during discussions and presentations, to foster a greater understanding among internal and external colleagues and constituents.*

In early-2014, to help coordinate overall GMI development with its separate products, priorities, and services, a project management approach was launched. Amy Esnard took the lead in developing a GMI project management work breakdown structure (GIS Management Institute. 2014c). This laid out the high-level tasks that correlated with the GMI Business Plan. They addressed both new product development and roll out, as well as ongoing GMI operations and product/service maintenance.

These tasks included:

Task 1: Project management

Task 2: Maintain core published GMI documents

Task 3: Develop and maintain the GMI technical infrastructure

Task 4: Update and maintain the GMI website

Task 5: Develop the 2014 GMI State of Organizational GIS Analysis Report

Task 6: Develop GMI GIS Assessment/Accreditation Subscription Service Pricing Schedule

Task 7: 2014 GMI GIS Assessment/Accreditation Subscription Service Soft Launch

Task 8: 2014 GMI GIS Assessment/Accreditation Subscription Service Full Launch

Allen Ibaugh agreed to load the work breakdown structure into MS Project, to facilitate management of GMI subcommittees which would be responsible for individual tasks and subtasks. This project plan was reviewed and adopted during the GMI Core Committee meeting on May 27, 2014.

By this point in mid-2014 A. Esnard had become very active on the GMI Committee. In January 2015 Esnard agreed to take over the GIS Management Institute Chair position. A GMI Technology Subcommittee was formed to design, develop, test, and pilot launch the GMI subscriber front end and database. The GMI Technology Subcommittee was chaired by A. Ibaugh and included A. Esnard, G. Babinski, K. Brennan, W. Nelson, and C. Cartee.

At the January 27, 2015 GMI Committee meeting, two significant actions were taken. This was the last GIS Management Institute Committee meeting that I chaired. By this time the GMI Committee had expanded to 27 members. Also attending this meeting were six URISA Board members: C. Anderson, A. Esnard, K. Mickey, A. Ibaugh, R. Somers, and D. Adams. Other attendees included D. DiBiase, D. Haley, N. Obermeyer, H. Perkins, B. Townsend, M. Vogt, and W. Nelson, URISA Executive Director.

The first action at this meeting was to approve 'the transition of the GMI Committee into the Core Committee of the GIS Management Institute as a URISA Operating Unit (ref 2015 URISA Bylaws sections 204 and 204.1.2).' This action was adopted unanimously by the GMI Committee. Second, the GIS Management Institute Core Committee endorsed A. Esnard as the new Chair.[xliii]

The Core Committee was established as the body that would coordinate GMI activity, establish subcommittees and task forces as needed, and plan and manage overall activity. The 2015 GMI Core Committee included:

- A. Esnard, Chair
- N. Obermeyer
- D. Haley
- R. Somers
- A. Ibaugh
- B. Townsend
- M. Vogt
- W. Nelson
- G. Babinski

The GMI Technology Subcommittee performed the bulk of the GMI work for the next eight months to prepare for a pilot launch of the GMI GIS Assessment Service by the end of 2015.

By this time the GMI GIS assessment application and database was in place and had been thoroughly tested. URISA selected five agencies to pilot test the full GMI GIS Assessment. Feedback from this pilot phase was used to further refine the GIS Assessment service. The GMI GIS Assessment Service was launched in early 2016.

Updated GMI Vision

At the same time, the GMI Core Committee updated the GIS Management Institute Conceptual Diagram to recognize the addition of the Metric Survey, Glossary, and focus on an initial GIS Assessment service as a precursor to the eventual GIS Accreditation Program (figure 5, S4).

Figure 8 illustrates the GMI implementation plan by October 2015.

In parallel with the development of the GMI GIS Assessment Service, the GMI launched a program to focus on publishing and promoting documents and papers of significance to the GIS management profession. In March 2014, a paper titled 'A Distributed Model for Effective National Geospatial Data Management: Building a National Data Sharing Infrastructure (Sparks et.al. 2014) was published as a GMI monograph. The following year a paper titled 'Report on National Survey of Multi-Organization GIS Programs' (Croswell 2015), was published as a second GMI monograph.

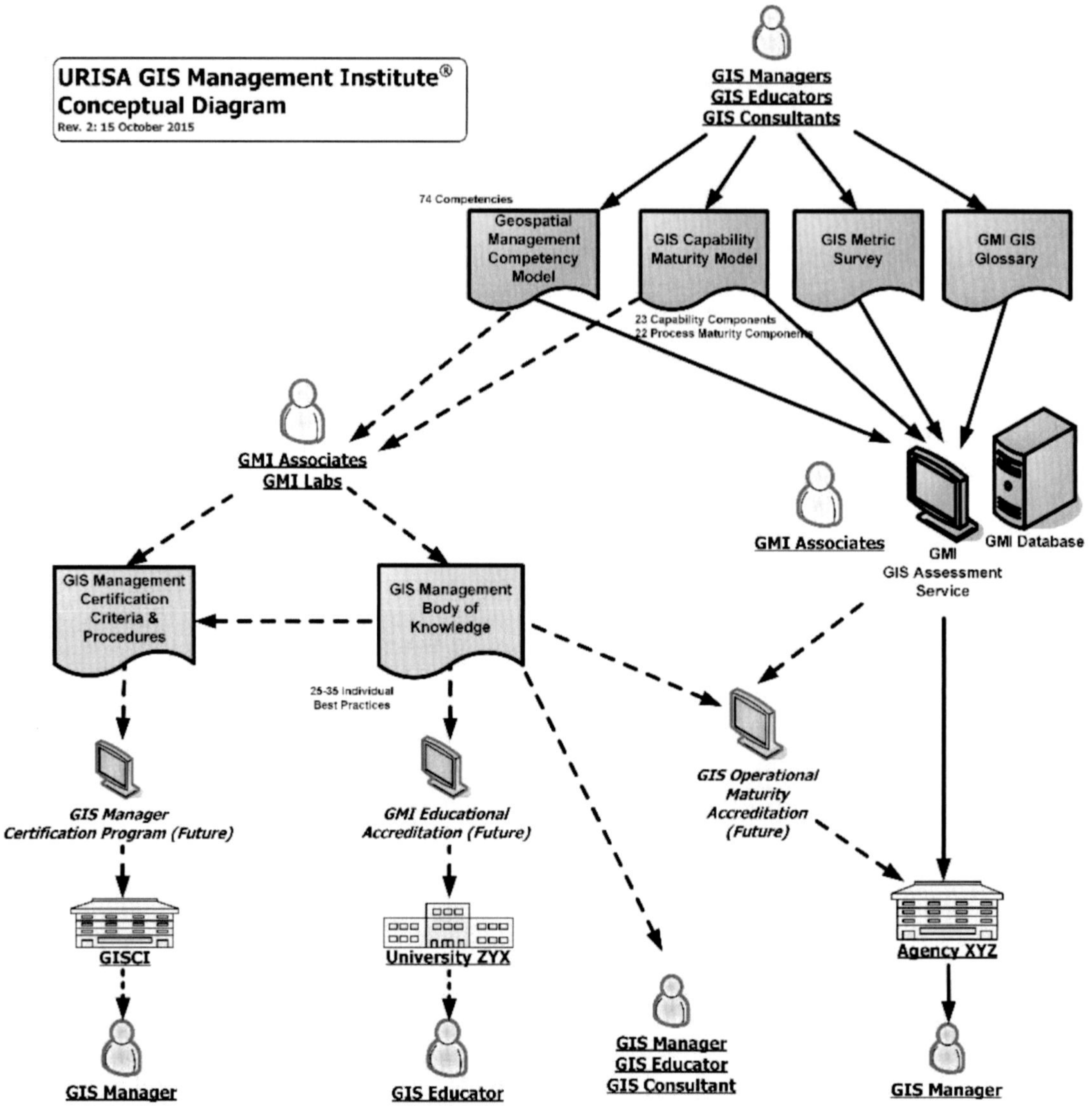

Figure 8
2015 GMI Conceptual Diagram

GMI Outreach 2014-2015

GMI outreach and education activities continued in parallel with GIS Management Institute development work.

In 2012 and 2014 discussions were held during the URISA Caribbean Conference with representatives of the International Cartographic Association (ICA) regarding high-level cooperation. In particular, ICA was interested in a mechanism to assess local government agency GIS operations that were funded by the European Union's INSPIRE program.[xliv] An MOU was signed in 2014 between URISA and ICA that provided for 'Promotion and awareness of the URISA GIS Management Institute GIS Organizational Assessment Service within the ICA community' (URISA-ICA 2014).

In early 2014 a half-day URISA Certified Workshop was developed titled: 'GIS Capability Maturity Model' (Babinski et.al. 2014). This workshop was designed to introduce the use of the GIS Capability Maturity Model for organizational assessment purposes. Through a hands-on exercise, workshop attendees complete an initial assessment of their own GIS operation. This workshop would be presented several times over the next year and a half. The goal was to familiarize GIS managers with the GISCMM and prepare them for participating in the future GMI GIS Assessment service.

Additional GIS Management Institute related outreach activity during 2014-2015 included:

- The GIS Capability Maturity Model workshop was previewed at a meeting of the Southeast Washington & Northeast Oregon GIS User Group on April 3, 2014 in Walla Walla, Washington (Babinski 2014b).
- In early 2014 a custom transportation-agency focused version of the half-day GISCMM workshop was prepared. This workshop was presented by Ibaugh and Butler at the 2014 GIS-T Conference in Burlington, Vermont (Butler et.al. 2014). At this conference, Federal Highway Administration (FHWA) and state DOT staff provided valuable input to GMI about the workshop and about the GMI GIS Assessment service.
- The GIS Capability Maturity Model workshop was presented at the Esri User Conference in San Diego on July 16, 2014. Also at the Esri UC a paper titled 'GIS Management Institute: The GIS Capability Maturity Model and the GMI GIS Assessment Service' was presented (Babinski 2014c).
- July-August 2014 *GIS Professional* - President's Column (Ibaugh 2014d) – mentioned continued GMI development within URISA.
- The GIS Capability Maturity Model workshop was again presented by J. Albrecht and Babinski at the 2014 GIS-Pro Conference in New Orleans on September 8, 2014. GMI figured prominently elsewhere in the conference program, being highlighted in the 'Welcome' message (Esnard 2014) and with a plenary session GMI status report on September 10 (Babinski, Ibaugh & Esnard 2014).
- On September 17, 2014 A. Esnard and I presented a URISA Connect GMI webinar titled: The GIS Capability Maturity Model: Maximize Benefits from Enterprise GIS Operations.' (Babinski and Esnard 2014)
- September-October 2014 *GIS Professional* - President's Column (Anderson 2014a) – made no mention of GMI. A URISA Past-President article (Smith & Ayan 2014) on GIS and three pillars of sustainability mentioned GMI; GIS-Pro 2014 URISA Committee Report (T. Townsend 2014) documented extensive GMI activity and planning.
- On November 4, 2014 a keynote address titled 'URISA, the Geospatial Revolution, and the GIS Management Institute' was presented at the 2014 New England URISA Day at Old Sturbridge Village, Massachusetts (Babinski 2014d).
- At the New England URISA event on November 4, 2014, I also met with representatives from South Ural State University, Chelyabinsk, Russia. A draft outline of a possible MOU between SUSU and URISA was prepared to facilitate the development of a Russian Chapter of URISA centered at SUSU. This would have included translation and publication of the URISA GISCMM, GMCM, and GMI

Glossary, and availability of the URISA GMI GIS Assessment service within Russia.[xlv]

- November-December 2014 *GIS Professional* - President's Column (Anderson 2014b) mentioned the GMI GIS Metric Survey; also, a small notice announced the GMI GIS Metrics Survey.

- January-February 2015 *GIS Professional* – there was no President's Column and no other mention of GMI.

- March-April 2015 *GIS Professional* again lacked a President's Column. This issue included an article titled 'Charting the Path Forward: URISA Directions for GIS Best Practices and Educational Development' (Peloquin and Bowe 2015) which described the work during the preceding year of URISA's Professional Practices Division and Education Division. It outlined material developed to '...support the GIS Management Institute as well as other URISA Initiatives.' This article mentioned the potential use of the GIS Capability Maturity Model to develop best practices and educational content, but without reference to the GMI Vision or proposed programs. However, there is no record in GMI Committee minutes that a report on the specifics of these Professional Practices Division and Education Division activities were communicated to GMI. This *GIS Professional* issue also included a small notice on the publication by GMI of a *Report on National Multi-Organizational GIS Programs.*

- An article about applying the GISCMM for public agencies titled 'Building Sustainable GIS Management with URISA's GIS Capability Maturity Model', was published in the Spring 2015 issue of *ArcNews* (Babinski and Esnard 2015a).

- May-June 2015 *GIS Professional* again lacked a President's Column. An article titled *'Is local government GIS supporting the vision of the NSDI?'* discussed the GMI GISCMM in terms of identifying if and how local government agencies '...track their progress towards supporting the vision of the NSDI.' (Ball 2015)

- On June 3, 2015 a paper about URISA and GMI was presented at a meeting of the Seattle Northwest Chapter of TDWI (The Data Warehouse Institute) in Seattle, Washington (Babinski 2015b).

- On September 29, 2015, at the 2015 ICMA Conference in Seattle, I participated on a GIS Management discussion panel. I spoke about the value of the GMI GIS Assessment for professional management of GIS (Babinski 2015c).

- At GIS-Pro 2015 in Spokane, Washington, Esnard and I presented a paper on October 21, 2015 titled 'Update on the URISA GIS Management Institute GIS Assessment Service' (Babinski and Esnard 2015b).

- On November 5, 2015, a webinar was presented for the Federal Highway Administration, GIS in Transportation Group titled: 'GIS Return on Investment' (Babinski 2015d). This paper included a focus on the link between GIS organizational process maturity, operational effectiveness and ROI.

On October 12, 2015, the King County GIS Center (Seattle Washington) issued a report titled: 'King County GIS Center 2015 GIS Assessment Report' (Babinski et.al. 2015). This report provided a comprehensive and self-critical assessment of the strengths and development and performance weakness of its program. The assessment was based on the URISA GMI GISCMM and GMCM.

In early 2016, without major announcement, URISA launched the GMI GIS Assessment service.

On March 22, 2016, I presented a paper promoting the new GMI GIS Assessment Service at the 2016 American Society of Public Administration in Seattle, Washington titled: 'Managing a County GIS for Results' (Babinski 2016).

In April 2016, King County subscribed to the URISA GMI GIS Assessment Service and completed its online assessment on April 19, 2016.

The GIS Community Takes Note

From 2009, when the GIS Capability Maturity Model was first developed, the GIS community has made note of GISCMM development by URISA and acknowledged the benefits of its application to GIS management practices.

'Model for Assessing GIS Maturity of an Organization' (Makela 2012) noted the usefulness of the URISA GISCMM to inform a proposed model for assessing GIS operations in Finland.

The City of Grand Rapids, Michigan Fourth Annual Information Technology and Change Management Report (Grand Rapids Executive Commission. 2012) stated: 'To ensure continued growth and smart investment the City's GIS Program has adopted the Urban and Regional Information System Association's GIS Capability Maturity Model. This model provides a method of self-assessment to ensure the City's GIS Program is supporting line of business strategies and helping employees develop effective, innovative learning models.'

Other local public agencies that utilized the GIS Capability Maturity Model to conduct their own self-assessment include the San Jose/Santa Clara Regional Wastewater Facility in California, Hennepin County, Minnesota, and King County, Washington (Babinski and Esnard 2015a). Many other agencies submitted completed self assessments to the GMI Committee with comments, requests for process clarification, and suggestions.

The San Jose/Santa Clara Regional Wastewater Facility Report (Hayes 2014) validated the process inherent in the GISCMM for documenting the justification for self-rating scores. The SJ/SCRWF Report used a copy of the published GISCMM to indicate its ratings for each component. It also included 53 pages of supplemental material to discuss the rationale for each individual self-rating, as well as organizational and technical diagrams, reports, and statements about how SJ/SCRWF would act upon its findings.

In 2012 COGO launched a project to develop a Geospatial Report Card for the Nation, similar to the 'Report Card for America's Infrastructure'[xlvi] published periodically by the Association of Civil Engineers. COGO's purpose and goal for the Geospatial Report Card for the Nation was described as:

> *The NSDI Report Card is a qualitative evaluation of the status and condition of the NSDI and its Framework data layers. It does not include cost estimates for completing the NSDI or for bringing the Framework to a specified level. The goal of this evaluation and report is to bring attention to the need for current and accurate geospatial data for the United States.*

As URISA's delegate to COGO, I provided input as a Report Card reviewer, including recommendations that future national geospatial assessments of the state of the NSDI would benefit by compiling and analyzing data from the NSGIG State Geospatial Maturity Assessment (NSGIC 2015) and from the GMI GIS Capability Maturity Assessment. The 'Report Card on the U.S. National Data Infrastructure' (Bossler et.al. 2015) was published by COGO on February 6, 2015. Commenting on the future value of the GMI GIS Capability Maturity Assessment, the report stated:

> *The URISA GIS Capability Maturity Model will serve as a stand-alone document to define the framework for an effective enterprise GIS. The model was developed initially with a focus on local government agencies (cities, counties, regional agencies, and similar entities), but it is intended for future use by any enterprise GIS. The model will help GIS managers and decision makers discuss the appropriate components of a capable enterprise GIS, the characteristics of a well managed GIS, and the effectiveness and ROI from a given level of investment. While it was not used in this assessment, it should prove valuable in future assessments.*

In late 2013 I was approached by a representative of the U.S. Department of Homeland Security, regarding a capability maturity model project for DHS processes.[xlvii] DHS interest focused on understanding the GISCMM, how it was planned to implement it as an assessment service within GMI, and the suitability of the CMM approach for complex Homeland Security-related geospatial data that needed to be accessible and accurate across multiple platform in real time. A subsequent report (Director of National Intelligence 2014) referred to the URISA GIS Capability Maturity Model:

> *The URISA Model is intended to serve the GIS community as a stand-alone document to define the framework for an effective enterprise GIS. The Model was developed initially with a focus on local government agencies (e.g., cities, counties, regional agencies, and other similar entities) but it is intended for future use by any enterprise GIS.'*

The information technology consulting service Gartrell Group published a report analyzing the GISCMM shortly after its publication by URISA (McLeod 2014). This report stated:

> *If used properly, the results of the survey can help an organization identify areas in need of improvement, and can be used as a metric going forward. Not only can the survey be used as a tool to measure individual organizations, but once many organizations have participated, their results can be combined to give a decent picture of the current state of GIS.*
>
> *One of the interesting elements of the assessment is a radar chart infographic that is used to depict the relative "maturity level" of an organization for each metric. A chart with circular results will show a fully robust organization without any need for improvement (let's be clear, a fully robust organization without any need for improvement is purely theoretical), while a chart with valleys cut into it will identify areas in need of improvement.*

'Evaluating Local Government Usage of GIS: A New Maturity Model' (Alrwais et.al 2015), provides a comprehensive comparison of past work on modelling GIS maturity, noting that the GISCMM has a different focus than other attempts:

> *'URISA's GIS maturity model focused on optimizing GIS operation (quality of GIS processes, enabling capability, and execution capability)...Although some of these models aim at describing maturity in using GIS, their benchmark variables (process area, critical success factors, and best practices) don't*

correspond tightly with GIS usage. Some of those models cover a wide range of perspectives that it's questionable to label it as a maturity model where in fact they are more of an IT/GIS management framework. The purpose of maturity models is to develop a simple yet comprehensive method of diagnosing an organization's maturity, but except for URISA none of the other models disclose the measurement tool.'

'Towards a Theory of GIS Program Management' (Albrecht 2015) states:

'Efforts to build a scientific foundation for [GIS] program management practices just started with the development of a thorough return on investment study across a multitude of organization types. Similarly, the data created by hundreds of organizations, who are as of 2015 conducting a GISCMM-based assessment will provide the foundation to build a generalized body of knowledge for GIS program management. Together with an analysis of best practices as identified, for example, through 35 years of peer review for the Exemplary Systems in Government Award, we are inching toward a theory of GIS Program Management.'

In 2015, a 'Survey on Maturity Models for the Geospatial Domain', (Karalopoulus and Kavouras 2015) stated 'Among them, the 'GIS Capability Maturity Model' focuses primarily on the GIS operations as such and it can be considered as the most extensive model, requiring in principle the involvement of GIS experts with deep and broad knowledge of the GIS environment inside the organization.'

A *GIS Professional* article titled 'Charting the Path Forward: URISA Directions for GIS Best Practices and Educational Development' (Peloquin and Bowe 2015) described 'activities designed to identify priorities within the geospatial community that can guide its work on the development of GIS best practices and educational offerings.' The article outlines activities of the URISA Professional Practices Division during the previous year. While this article recognized...

'...the identification of existing best practices that relate to existing recognized models, such as the Geographic Information Science and Technology (GIS&T) Body of Knowledge, the Geospatial Management Competency Model (GMCM) and the GIS Capability Maturity Model (GISCMM), that describe knowledge areas and associated skills that characterize GIS professionals.'

However, the article does not indicate how so-called 'best practices' related to GIS&T, GMCM, and GISCMM were identified or how they were validated as 'best practices.'

A recent book, 'Advancing Geographic Information Science: The Past and Next Twenty Years' (Onsrud and Kuhn 2016) comments on the need to mature the study of GIS management practices. It cites the GISCMM and GISCMM-based assessments as a potential future '...foundation to build a generalized body of knowledge for GIS Program management.'

Most recently URISA's GMI activity received attention in India, for example, 'A View on GIS Capability Maturity Model (GISCMM)' (IndraStar 2016).

A recent dissertation (Alrwais 2016) notes that of twelve GIS organizational maturity models studied, only the URISA GMI GISCMM discloses its measurement tool and is focused on optimizing the quality of GIS related processes.

Continuing GMI Development

The GIS Management Institute Vision is far from being achieved. The key items from the original Vision that should be addressed by GMI and URISA include:

- Activate a professional practices initiative within URISA to develop the GIS Management Body of Knowledge (P3, above). Individual Best Practices should be compiled for each component of the GISCMM. In 2012 Carl Anderson proposed a 'URISA Labs' process to accomplish this task. For each GISCMM topic, a URISA Lab would be mobilized to bring together qualified and interested GIS professionals. They would serve as a research team to compile various real-world and academic practices related to the topic, draft a proposed 'best practice', and submit it for public peer-review. After peer review, the URISA Lab would revise the draft as appropriate and publish the GMI Best Practice under their authorship. This work would be facilitated by the current GMI GIS Assessment Service, which requires subscribes to submit policies and procedures for EA (Execution Ability or process maturity) components rated at Level 3 or above. These policies and procedures would be sorted by the corresponding EA categories as real-world examples of current practices. Thus, they would provide a resource for all agencies to develop or refine their individual best-practice development.
- Once the GMBOK is in development, begin work on the components of the proposed GIS Accreditation Service for organizations (S3, above) and Accreditation Service for GIS educational programs (S6, above).
- Build a GMI Knowledge Base web-site for GMI GIS Assessment subscribers where they could post copies of their GIS related policies, procedures, practices, standards, and metrics.[xlviii] Documents would be sharable between agencies, and usable for research by URISA Labs researchers. This resource would be valuable for agencies before GMBOK best practices are developed. Even after GMBOK best practices are published, the GMI Knowledge Base web-site would be valuable for tracking the implementation and use of GMBOK best practices and facilitate periodic future GMBOK updates and revisions as real-world practices evolve and improve further.
- Return on investment is a very important factor for public agencies to secure adequate GIS funding. The 2012 King County ROI study (Zerbe et.al 2016) was significant in that for the first time an ROI model was applied to GIS by a non-GIS research group. The importance of a credible ROI methodology to the broad GIS has been frequently noted (Smith 2011b). A large GIS ROI study has been proposed to determine the relationship between an organization's GIS capability and maturity and the financial return on investment (ROI) generated for the organization (Babinski 2016). The objectives of this proposed study would include:
  - Study 30 municipalities (study agencies) in the U.S. and Canada. A minimum of 30 study agencies would be sufficient for statistically valid analysis of results.
  - For each study agency, assess their organizational metrics, GIS capability, GIS maturity, and geospatial management competency against the GMI GIS Assessment Service
  - For each study agency apply the Zerbe GIS ROI methodology to determine their ROI

- Analyze and report on statistically valid correlations between individual GIS capability and maturity components with ROI.

8
URISA ELIMINATES THE GMI VISION AND REDUCES THE SCOPE

In mid-2015 the URISA Board unexpectedly and without explanation eliminated its support for the previously adopted GMI Vision and Scope. It took this action without consultation with the GMI Core Committee. Since that time, the nature and level of URISA Board support for the long-term development of GMI has not been openly stated. There is no published URISA vision for GMI.

In October 2014, the new URISA board that first convened at GIS-Pro in New Orleans decided to conduct a review of its Bylaws. At Board request, a small committee was formed of URISA Past-Presidents, chaired by a current URISA Board member. I volunteered to serve on this Core Documents Committee.[xlix] The Core Documents Committee worked into mid-2015. Several of its final recommendations directly affected the GIS Management Institute:

- Eliminate the Operating Division Structure from the March 2013 Bylaws, including elimination of the GIS Management Institute as an operating unit.
- Move the Professional Practices Division, which was responsible for developing the GMBOK, out of GMI.
- Eliminate all committees from the bylaws except for an Executive Committee, Finance Committee, and Strategic Planning Committee.
- Move all other committees, including GMI, from the bylaws to the URISA policy Manual
- Reduce the scope of the GMI Committee, eliminating most of what was approved by the URISA Board in the March 2013 bylaws.

As a Core Documents Committee member, I opposed these changes. I proposed an alternate description of the GMI Committee for consideration by the Core Documents Committee that would have preserved the essence of the Vision for the GIS Management Institute in the Policy Manual as it had been adopted by the GMI Committee and approved by the URISA Board in the March 2013 Bylaws. This alternate GMI description that I proposed was rejected by the Core Documents Committee.[l]

During its July 2015 face-to-face meeting, the URISA Board approved the revised URISA Bylaws and Policy Manual as proposed by the Core Documents Committee. There was no attempt by the 2015 URISA Board, prior to this decision, to consult past URISA Board members and GMI Committee members who were involved in developing the GIS Management Institute, drafting the 2013 URISA Bylaws, and defining the GMI Scope and Vision.

The chair of the Core Documents Committee reported to me on the action that the URISA Board took at its July 2015 meeting:

> *The Board concurred with the recommendations of the Core Docs committee about the wording to include in the Policy Manual. However, the Board also unanimously agreed that a letter should be send [sic] from the President to the GMI that establishes the strong support of the Board for their work and that encourages them to pursue the vision that they have defined. As GMI seeks to expand its*

> *services they will be asked to submit a plan to the Board as soon as practical for achieving each of the goals that they have defined. Future versions of the Policy Manual will be updated in a timely manner to reflect the current status of GMI as it evolves.*[li]

Through the summer and early autumn of 2015 no letter of support from the URISA President was received by the GMI Committee. I wrote to the then URISA President, Past-President, and President-Elect on November 10, 2015 inquiring about the level of URISA Board support for GMI.[lii] To date I have received no reply.

It is of course the right of the URISA Board of Directors to take any action that it believes is in the best interests of the Association. But in this case, the Board's action, taken without prior consultation with the full GMI Committee and with past URISA Board members, the lack of subsequent explanation regarding the elimination of the previously adopted GMI Scope and Vision, and the lack of a statement of support for the GMI Vision, all bring into question the URISA Board of Director's support of and commitment to a program that had been more than seven years in development by a large team of committed GIS professionals and URISA volunteers. For more than seven years, scores of volunteers participated in hundreds of meetings and conference calls, and devoted many thousands of hours of effort to developing the GIS Management Institute Scope and Vision.

The existence of revised Bylaws and Policy Manual were announced at GIS-pro 2015 in Spokane, but it was only in April 2016 that the Bylaws appeared on the URISA website. An unattributed article in the September-October 2015 *GIS Professional* titled 'URISA Adopts New and Improved Organizational Structure' stated that ...within the Policy Manual, each committee's Purpose and Scope, Membership and Chairmanship, and Committee-specific Policies and Procedures are detailed.'[liii] But as of until late-2017 the URISA Policy Manual could not be found on the Association website.[liv]

The URISA Bylaws now state: '304.4 Other Committees: Other committees are defined in the Policy Manual.'[lv] An undated document on the URISA website titled 'URISA Committees' states:

> *The following committees are described within the URISA Policy Manual, adopted October 2015. Each committee's purpose and scope; membership and chairmanship; and committee-specific policies and procedures are included.*[lvi]
>
> *GIS Management Institute (GMI) Committee*
>
> *Purpose and Scope. The GIS Management Institute (GMI) develops, manages, and implements professional GIS management products, and services, necessary to assess the maturity and capability of GIS operations. It coordinates administrative activities that support, market, and provide GMI products and services as performed by URISA headquarters staff and volunteers.*
>
> *Membership and Chairmanship. Open to all URISA members.*
>
> *Committee Specific Policies and Procedures.*
>
> *The GMI Committee provides a representative (Chair or other representative) to be a member of the Professional Educational Committee (PEMC), and that representative participates in the semi-annual professional education materials*

coordination meetings convened by the PEMC. Each month the GMI representative transmits a list of professional education activities to the other committees represented on the PEMC.

Shortly after the 2015 GIS-Pro Conference, without explanation or consultation with the GMI Core Committee, the URISA Board replaced A. Esnard with A. Ibaugh as GMI Committee Chair[lvii]. Later, again without consultation with the GMI Committee, Ibaugh was replaced as GMI Committee Chair in 2017 by R. Somers. The last GMI Committee meeting occurred in the first half of 2016.

Why did the URISA Board change its level of support for GMI without explanation or consultation with the GMI Committee? Why did it eliminate the GMI Vision? Some of my colleagues have speculated that academics on the URISA Board felt threatened by the proposal to develop the GMBOK outside the academic environment. However, respected academics participated on the GMI Committee and on the URISA Board and never voiced open objections to the GMI Vision and Plan. Others have suggested that GIS management consultants on the GMI Committee or on the URISA Board felt threatened by GMI and the tools and capabilities that it would provide for agencies and companies that operate GIS, thus threatening a part of their business. However, several prominent GIS consultants participated on the GMI Committee and on the URISA Board of Directors and never openly voiced objections to the GMI Vision and Plan.

Absent an explanation or a statement of GMI support by URISA, all that I and the GIS community can do is to speculate.

9
THE FUTURE OF THE GIS MANAGEMENT INSTITUTE

The GMI Committee demonstrated the financial viability of the GMI products and services portfolio as outlined in the Business Plan and Financial Plan. For several years the URISA Board of Directors formally expressed strong support for the GMI Vision and Business Plan.

After URISA eliminated to previously approved GIS Management Institute Vision and Scope in mid-2015, there was sufficient GMI momentum to complete the final development and launch of the GMI GIS Assessment Service in mid-2015.

There were face-to-face GMI Committee meetings at the URISA GIS-Pro Conferences in 2016 and 2017, however no minutes were recorded of these meetings. As of April 2018, the last regular monthly URISA GMI Committee meeting had occurred in March 2016. In the six months before that there were only two other GMI Committee meetings, in September 2015 and February 2016. A meeting scheduled for April 2016 was cancelled and no other meetings have been scheduled. The URISA website, as of April 2018, still shows a GMI Committee on its organization chart and URISA's leadership information lists a GIS Management Institute Chair and Board of Directors Liaison to GMI.[lviii]

At GIS-Pro 2016 in Toronto, there was minimal focus given to GMI. The URISA GISCMM half-day workshop was presented during the pre-conference. One 90-minute session titled 'The Importance of GIS Program Review and Assessment for Successful GIS Managers' included a paper that focused on the GMI GIS Assessment Service (Esnard, Ibaugh, B. Townsend, 2016).

At GIS-Pro 2017 in Jacksonville, Florida, the GISCMM half-day workshop was dropped from the program. The only GMI-related content was a 90-minute session titled 'GIS Program Review & Assessment: Where You Are and How to Get Where You Want to Be.' Three papers addressed GIS program review and assessment methodology (Somers 2017), how to utilize the GISCMM at the local agency level (B. Townsend 2017), and the use of the GISCMM at State DOTs (Ibaugh 2017).

Lacking a statement of support for the GIS Management Institute from the URISA Board of Directors or from the URISA President since the time of the mid-2015 Bylaws and Policy Manual changes, how can the level of URISA leadership support be gauged? A high-level review of the topics of President's Columns in URISA's bi-monthly *GIS Professional* newsletter, after the 2012 URISA press release announcing the formation of GMI, provides some evidence.

- In 2013, all six President's Columns referred to GMI.
- In 2014, five of six President's Columns referred to GMI.
- In 2015, only one President's Columns referred to GMI.
- In 2016, three President's Columns referred to GMI.
- In 2017, only two President's Columns referred to GMI.

Clearly, for URISA leadership, GMI is not forgotten, but it has hardly seemed a leadership priority during the last three years.

A review of references to GMI in the *GIS Professional* since mid-2015 confirms uneven interest and promotion by URISA and URISA leadership.

- July-August 2015 *GIS professional* had no President's Column and no other mention of GMI.
- September-October 2015 *GIS Professional* again had no President's Column. An article in this issue titled 'URISA Adopts New and Improved Organizational Structure' outlined the work of the URISA Core Documents Committee that recommended the elimination of the GMI Vision as outlined above. The article presented a URISA Organization Chart that showed GMI as a URISA Committee.[lix] But it did not explain why or how the organizational structure was 'improved'. It made no reference to the process and did not reference the lack of consultation with the GMI Committee.
- November-December 2015 *GIS Professional* - President's column (Somers 2015) mentioned an assumed 2016 expansion of GMI, in addition to other URISA activities.
- January-February 2016 *GIS Professional.* This issue had no President's Column and no other mention of GMI. This issue included a call for papers for the 2016 GIS-Pro Conference, listing many desired topic areas, but not including any related to GMI, the GISCMM, or GMCM.
- March-April 2016 *GIS Professional* - The President's Column (Somers 2016a) focused on 'GIS Program Review and Assessment.' Among various assessment resources, it referenced the expected 2016 launch of a GMI 'GIS Capability Maturity Model Assessment Service.' It included a link to the URISA GMI website.
- May-June 2016 *GIS Professional* – President's Column (Somers 2016b) focused on URISA support to the GIS profession. It mentioned best-practice guides being developed by URISA without reference to GMI or to the GISCMM. There was no mention of GMI, GISCMM, or GMCM. This issue also included a two-page article about the planned 2016 URISA GIS-Pro Conference, but with no reference to any planned GMI-specific activity.
- July-August 2016 *GIS Professional* - President's Column (Somers 2016c) focused on the need to manage a GIS to defined best practices. The article stated that URISA '....has been working on developing a series of GIS best practices to support...general needs, and in particular the GMI GIS Capability Maturity Model assessment.' The article included the GIS management best practice development process and listed (without reference or attribution to GMI) many of the proposed best practice components that were first documented in the GMI Business Plan (Babinski 2013k). There was no reference to GMI activity or a call for GMI volunteers in this article.
- September-October 2016 *GIS Professional* - President's Column (Somers 2016d) focused on URISA volunteer opportunities and needs. It referred to GMI:

> *The GIS Management Institute (GMI) develops resources to help organizations assess and improve the capability and maturity of their GIS operations. It has recently expanded the GIS Capability Maturity Model (GISCMM) participation to several new organizations and continues to grow and seek new agency participation.*

But this article did not refer to a GMI Committee, implying that the GIS Management Institute is simply an operational component of URISA, without the need for a cadre of GIS professional volunteers.

- November-December 2016 *GIS professional* – President's Column (Corbin 2016) referred to URISA's newly revised bylaws and operating manual as means '...to provide better and increasing services...' to members. However, there was no mention of GMI in this column or anywhere else in this issue.
- January-February 2017 *GIS Professional* – President's Column (Corbin 2017) made no mention of GMI or of any GMI-related activity. This issue included a call for presentations for the 2017 GIS-Pro Conference. One suggested topic was 'GIS Capability Maturity Model' but it was not clear if this referred to the URISA GISCMM or some other model.[lx]

The GIS Community Continues to Take Note of GMI

After the URISA Bylaws and Policy Manual changes and the elimination of the GMI Vision, 2016 was a year of little activity and URISA focus related to the GIS Management Institute. However, outside URISA, the wider GIS community continued to see value in GIS operational maturity assessment and GIS management best practices.

Analysis of the characteristics and effectiveness of GIS in local and state government operations continues to be a topic of professional interest and study. One recent study analyzed how GIS professionals put GIS to work for the agencies that employ them in the Carolinas. This study (Bradbard et.al, 2017) looked at GIS from a variety of roles within a typical municipal operation, including 1) use in routine operations versus decision support, 2) the perceived impact and level of GIS success, 3) Methods of implementing GIS projects, and 4) the role of GIS in eGovernment and its impact on privacy issues.

The analysis was based on a custom designed three-part research instrument: 1) a written questionnaire, 2) a structured face to face survey instrument, and 3) a site visit check-list. The authors demonstrated that understanding and measuring the impact of GIS on an organization is an important component for managers and stakeholders for future planning, refinement, resource allocation, and operations.

The 2015-2016 King County GIS Center Management Team assessment report referred to above was published in the *URISA Journal* (Babinski et.al, 2017). This was the first public reporting of a detailed and comprehensive analysis of a local government GIS operation against the GISCMM, GMCM, and the URISA Metrics Survey.

In 2016 and 2017 the Federal Highway Administration (FHWA), GIS in Transportation Group published two reports on the application of GIS maturity models for assessing the effectiveness and capabilities of state highway department GIS operations. This activity developed out of the GIS Capability Maturity Workshop (Butler et.al. 2014) presented at the GIS-T Symposium, Burlington, Vermont, May 5, 2014 by A. Ibaugh and A. Butler.

The first of these FHWA reports, 'GIS Capability Maturity Model Peer Exchange' (Samiento et.al. 2016) stated:

'This report provides highlights from a peer exchange held in Nashville, TN, on September 20-21, 2016. The exchange was held as part of the Federal Highway

Administration's (FHWA) Geographic Information Systems (GIS) in Transportation program and was hosted by the Tennessee Department of Transportation (TDOT). The purpose of the exchange was to highlight the role that organizational assessments can play in developing a comprehensive GIS strategy, by sharing experiences between State DOTs that have undergone organizational assessments and State DOTs that have limited experience with, or have not undergone, an assessment.'

The second FHWA report, 'Geospatial Information System Capability Maturity Models: Case Studies of Select Transportation Agencies (Green et.al. 2017) stated:

This report explores how State Departments of Transportation (DOTs) are assessing organizational GIS decision-making capabilities, particularly through the use of capability maturity models. This report supports GIS practitioners and decision-makers by identifying examples of noteworthy practices, considering the benefits and limitations of implementing a capability maturity model, and determining how to identify and address factors that alter the effectiveness of a GIS division in applying geospatial tools and services.

These two FHWA reports included useful, critical assessment of the use of the GISCMM for large state transportation agencies. These criticisms provide feedback that can be used for a future revision of the GIS Capability Maturity Model.

The GISCMM and GMCM have also been used in unanticipated ways. In 2017 the King County GIS Center in Seattle, Washington, began a project to integrate its services and operations into the King County Information Technology Department (KCIT). Prior to 2017, King County GIS had been in operation for 20 years. For the previous 16 years it had operated as a county internal service fund, with its own budget, finances, and organizational structure. As KCIT matured with an ITIL-based services-oriented organizational structure, King County decided to fully integrate King County GIS within the KCIT structure.

A consultant was retained to help facilitate the transition of King County GIS into the distributed KCIT-service service-oriented structure. KCIT and KCGIS management required the consultant to use the 2015 King County GIS Assessment Report (Babinski et al, 2017) to analyze the potential impact of any proposed change to the KCGIS organizational structure against the 2015 (and a more recent 2017) KCGIS GISCMM and GMCM GIS Assessment ratings. For each GISCMM and GMCM component, the KCGIS Reorganization report (Fox 2018) reported on the consensus impact of the proposed future organizational structure for each GISCMM and most GMCM components. This report also included detailed process flow diagrams to document King County GIS Center's pre-reorganization GIS business processes. This filled in the gaps of many KCGIS processes that could not be rated within the GISCMM as higher than maturity level 2 (repeatable) because no process documentation existed. These new process flow diagrams will also be used to document the future revised processes in the new service-oriented structure within KCIT.

GIS Management Institute 2017-2018

The GIS Management Institute, the GISCMM, and other GMI activities and products were not mentioned in any *GIS Professional* President's Column from November-December 2016 through the July-August 2017 issue.

In mid-2017 however, there were a few small hopeful signs within URISA regarding GMI. A series of four articles were published in *The GIS Professional* about GMI and the GIS Assessment Service. They included some verbatim content that the GMI Committee had authored about the GIS Management Institute years earlier. They also represented some capabilities of the GIS Assessment services that are not in fact available. For example, the ability for GMI subscribers who complete their GIS Assessment to compare their results with other agencies did not in fact seem possible.

The March-April 2017 *GIS Professional* GMI article was titled 'URISA's GIS Management Institute: What Can GMI Do For Me?' (Hitt & Townsend 2017a). This article presented a case for the benefit to GIS managers and key agency stakeholders to use GMI products, with focus on the GIS Capability Maturity Model. The article demonstrated a confused understanding of the term 'maturity' as used with the GISCMM. It failed to explain the two key components of the GISCMM: Enabling Capability and Execution Ability. It applied the term 'maturity' to the model as a whole, which failed to distinguish the GISCMM concept that a GIS operation can be assessed with both a resource (capability) focus and a process (maturity) focus with this GMI tool. The article failed to mention the GMI Committee and did not invite interested GIS professionals to volunteer to assist with GMI activities. It failed to include the URISA GMI web page URL, to assist discovering more information. The URISA President's Column in this issue made no mention of GMI.[lxi]

The May-June 2017 *GIS Professional* GMI article was titled 'Getting started with the GISCMM in 3 quick steps' (Hitt & Townsend 2017b). This article referenced the two options for using GMI GIS Assessment resources: either online via the GMI subscription service, or off line, by using the GISCMM and GMCM to complete an internal report. The article listed the GMI GIS Metric Survey as a component of the subscription service, but without explaining the purpose or value of the survey. This article also used confused terminology, variously referring to the GIS Capability Maturity Model as either the GISCMM or CMM. It introduced the Geospatial Management Competency Model as a component of GMI without explain what the GMCM is and how it relates to the GISCMM. It refers to the evaluation of results via spider diagrams, without clarifying that this is a component of the GMI GIS Assessment Service. Once again, this article failed to mention the GMI Committee and did not invite interested GIS professionals to volunteer to assist with GMI activities. It failed to include the URISA GMI web page URL, to assist discovering more information. The URISA President's Column in this issue made no mention of GMI.[lxii]

The July-August 2017 *GIS Professional* GMI article was titled: 'URISA's GIS Management Institute: What Happens After the Assessment?' (Hitt & Townsend 2017c). This article referred to the ability to '....compare your agency results to the results of an average of agencies with comparable characteristics through the subscription service.' In fact the ability to compare a subscriber agency results against peer agencies is not currently possible.[lxiii] The article also invited agencies to '...write a short description [of best practices] and submit it to the URISA GMI for inclusion in the best practices.' However, the article did not explain what the GMI or URISA 'best practices' will be or how they will be validated. There was no reference to the GMI goal of developing peer-reviewed GIS Management Best Practices into a GIS Management Body of Knowledge (GMBOK). Once again, the URISA President's Column in this issue made no mention of GMI.[lxiv]

The August-September 2017 *GIS Professional* GMI article was titled 'URISA's GIS Management Institute: Bringing It All Together (Hitt & Townsend 2017d). This article provided a summary of the three previous GMI articles. It also added comments from four public agencies that had used the GMI GIS Assessment service that reinforced the self-assessment process and explained some of the benefits. This article did not include a link to the GMI web site or information on how to learn more about GMI. The President's Column in this issue referred to the series of four GMI articles in the GIS Professional. It also mentioned that GMI had been working with the 'Transportation Industry' and state and local DOTs without elaborating about the nature of this work.[lxv]

Despite ignoring previous GMI development and the GMI Vision, it is encouraging that these articles present the GIS Management Institute and the GIS Assessment Service as programs about which URISA wants to inform the GIS community.

In the November-December 2017 *GIS Professional,* the President's column (T. Townsend 2017) outlined recent URISA accomplishments and strategic plan priorities for the coming year. There was no discussion about the Vision or current/planned activity of the GIS Management Institute. Under a heading 'Professional Practice and Educational Resources,' GMI was incorrectly referred to as the 'Geospatial Management Institute' with the Geospatial Management Competency Model and the 'GIS Capability Maturity Model and Assessment' as seemingly separate things. In referring to URISA's focus areas and strategic outcomes, the article lists under the heading 'Promote, protect and grow the GIS Profession' the goal to 'Increase GMI value and use of the GIS Capability Maturity Assessment tool.'

As of April 2018, the URISA website lists a GMI Committee Chair and a designate Board of Directors Liaison to GMI[lxvi]. But there has still been no recent GMI Committee meeting or call for new GMI volunteers.

On February 16, 2018, the GMI Committee Chair sent an email to about 20 people who had previously been GMI Committee members. This email stated that the GMI Committee had not meet at all the previous year. The email referred to 'GMI Goals in the Board's Strategic Plan' and that the GMI Committee would be "re-instituted' soon to work towards these goals. There was no indication however of what the URISA Board's goals might be.[lxvii]

At about the same time, the URISA GMI home page added a small section that outlined new "GMI Initiatives."[lxviii] These new initiatives were described as:

GMI is working on several more initiatives to help organizations with their GIS management operations. These include:

- *Development of GIS Management Best Practices and Body of Knowledge*
- *GIS Management-related professional education*
- *Expansion of GIS Management information dissemination*
- *Promotion of awareness and adoption of GIS management best practices by organizations and GIS professionals*

It is a positive sign that URISA has identified new GMI Initiatives and plans new GMI-specific goals for a newly reformed GMI Committee to pursue.

One Way Forward for the GIS Management Institute

URISA has never formally recognized any of the committees, subcommittees, and their members who worked over seven years to develop the GMI, GISCMM, GMCM, GMI Glossary, and the GMI GIS Assessment Service. Between 2009 and 2016, I have identified 50 individuals who contributed at various times to develop the GISCMM, GMCM, and the GIS Management Institute (figure 9).

The individual accomplishments of the GIS Management Institute are numerous. They include:

- Refinement and publication of the first comprehensive open peer-reviewed model to assess both the capability and the process maturity of GIS operations.
- The first open peer-reviewed rating scales for the assessment of GIS operational capability and maturity.
- The first open peer-reviewed geospatial management competency model.
- The first open rating scale to assess geospatial management competency.
- The development of an open, peer-reviewed GIS glossary specifically focused on the need to assess GIS operational capability and maturity, along with geospatial management competency, on a consistent basis.
- The development of the first open, peer-reviewed GIS operational metric survey.
- The launch of the first open comprehensive GIS-assessment and reporting service.
- The development of an organizational framework for the future development and compilation of peer-reviewed GIS best-practices, and a future GIS management body of knowledge.
- A proposal for the scientific study of the correlation between GIS capability and process maturity with return on investment.
- A proposal and conceptual mechanism for the accreditation of GIS operations and GIS management educational institutions.
- A proposal, conceptual mechanism, and agreement for the development of a GIS Manager Certification Program in cooperation with the GIS Certification Institute.
- A comprehensive vision and business plan for a self-sustaining organization that would support professional GIS management and operations anywhere.

The GIS Assessment Service launched in 2016 will require *consistent*, year after year URISA Board support, strong executive leadership, active and committed volunteers, led by a strong GMI Committee leader, and focused priority effort by all. Flaws in the GIS Assessment Service need to be quickly identified and corrected.

My personal observation is that most existing URISA activities are well defined and can succeed by applying program templates that have been used many times in the past. This includes programs like the annual GIS-Pro Conference and various specialty URISA conferences, educational content like URISA workshops and the URISA GIS Leadership Academy, and publications like the *URISA Journal* and the *GIS Professional.*

But GMI is different. It is new. Without a vision and development plan, and without strong, consistent support and leadership, it will fail. The GIS Management Institute will succeed if its Vision and Programs are supported by the Board and if the GMI Core Committee is given responsibility to execute its Vision. A parallel can be found in GISCorps, which has a well-defined vision and mission and is left to develop and manage its program with a large degree of autonomy and self-financing.

URISA created an ambiguous situation when it eliminated the GMI Vision and organizational structure. The URISA Board, despite request, has failed to state its vision and level of support for the GIS Management Institute to the GMI Committee and to the GIS professional community.

If URISA desires to pursue the original GMI Vision, key priorities include:

- A revitalized GMI Core Committee. Hopefully this will be implemented early in 2018.
- Effective and focused marketing of GMI products and services. Ideally this would occur within GMI, but if URISA wishes to utilize marketing resources outside GMI, this needs to be coordinated closely by GMI leadership. This coordination is necessary to ensure clear and consistent GMI products and services definitions and understanding of GMI business value for subscriber agencies.
- Effective support of the GMI GIS Assessment Service so that GMI subscribers achieve real and ongoing value for their GIS operations and management. Subscription fees should begin to generate financial resources so that the GMI program and its future development and expansion can become self-supporting.
- Development of the GIS Management Body of Knowledge, as conceived within the GMI Vision has not materialized. Mobilizing resources to develop the GMBOK *within GMI* would ensure steady and focused progress on this key resource.
- The GMI GIS Assessment Service should form the basis for a GMI GIS Accreditation program. This does not require completion of the GMBOK. Accreditation would be based on agency assessment against the GISCMM. In fact, a GIS Accreditation program would accelerate development of the GMBOK by identifying the practices that mature GIS operations actually use.
- GMI should plan for the eventual and ongoing review and updating of the GISCMM and GMCM (and eventually the GMBOK as well).
- There have been suggestions that a scaled-down version of the GISCMM should be developed to assess GIS operations for small agencies. This concept was considered several times during initial development and final review and revision of the model. However, the concept of a scaled down GISCMM was rejected, because all the components of the model can and should be applied to any assessment. It is likely that many small and medium sized agencies might struggle to achieve high capability ratings and a high level of process maturity. But this does not mean that they should not assess themselves against a single unified model. This will allow small and medium sized agencies that do achieve a high level of capability and process maturity to serve as examples for peer agencies.
- As the GMBOK is developed, a by-product will be the specifics of GIS management competency that can be used in partnership with GISCI to develop a GIS Management Certification program.
- Funding should be pursued for the proposed multi-agency GISCMM/ROI study. Results of this study would have the potential to put GIS operations and management on a solid scientific basis.

Curiously, the original GMI Vision is still stated in URISA's published GISCMM. Why is there this discrepancy between URISA policy and a published and peer-reviewed URISA document?

The purpose and scope of the GIS Management Institute as currently defined by URISA falls far short of the vision and planning that began in 2009. If the URISA

Board does not support this Vision, it need do nothing. It is the author's opinion that the URISA Board has created an ambiguous situation regarding support for GMI. If the Board supports GMI's Vision and is committed to its success, it needs to demonstrate support of the GIS Management Institute.

To 'Respect the work of others'[lxix] - the position of URISA regarding the past, present, and future of the GIS Management Institute, and the many volunteers who created it, should be clarified.

Noted architect and urban planner Daniel Hudson Burnham said, '...make no little plans.' A key risk-factor facing GMI is the fact that it is a very big audacious goal which requires a clear vision, a well-documented business plan, an ongoing level of understanding of that plan, and long-term support. Many of those who began discussing in 2009 what became the GIS Management Institute today are still involved. But, each year one-third of the URISA Board changes. If the GMI Vision is not sustained, supported, and carried forward by the URISA Board year after year, it will fail. The professional GIS community would be the less for that failure.

URISA GIS Management Institute		2008	2009	2009	2010	2010	2011	2011	2011 2012	2012 2013	2013	2013 2014	2015	2012	2013	2014	2015	2015
Volunteer Names		Original GISCMM Development	King County GISCMM Peer Review	URISA GISCMM AD-hoc Committee	URISA GIS-Pro GISCMM Workshop	GMCM Core Committee	Washington GIS Managers Work Session	GMI Committee	GMCM Task Force	GISCMM Task Force	GMI GIS Glossary Sub-Committee	GMI GIS Metrics Survey Sub-Committee	GIS Assessment Technical Sub-Committee	GMI Committee	GMI Committee	GMI Committee	GMI Committee	GMI Core Committee
Babinski	G	X	C	X	X	C	C	C	C	C	X	X	X	C	C	C	XC	X
Horning	G		X				X											
Castle	L		X															
Jacobson	M		X															
Gelb	R		X															
Perkins	H			C	X					X	X			X	X	X	X	
DiBiase	D			X		C	r	X	C	X				X	X	X	X	
McDermott	T			X	X				X									
Kennely	P					X			C									
Ayan	D				X					r								
Burdick	D						X											
Esnard	A						X			r			X			X	XC	XC
Von Essen	I						X			r								
Griffin	T						X											
Beimburn	S						X											
Almvig	G						r											
Smith	C							X		X				X	X	X	X	
Obermayer	N							X		X	C	X		X	X	X	X	X
Brown	C							X		X				X	X	X	X	
Conry	T								X									
Croswell	P								X							X	X	
Haley	D								X	X	X			X	X	X	X	X
Ryan	B								X									
Somers	R								X	X	X			X	X	X	X	X
Stauffer	B								X									
Sun	C								X					X	X	X		
Albrecht	J									X	C			X	X	X		
Anderson	C									X		X		X	X	X	X	
Bera	U									X					X	X	X	
Bohard	E									X				X	X	X	X	
Butler	A									X				X	X	X		
Eichelberger	P									X		X			X	X	X	
Mickey	K									X	X				X	X	X	
Roche	M									X				X	X	X	X	
Grant	V									X				X	X	X	X	
Ibaugh	A									r			C				X	XC
Terry	G									r								
Nelson	W									X			X	X	X	X	X	X
Brennan	K									X			X	X	X	X	X	
Cartee	C												X					
Al-Zifin	M													X	X			
Baldock	P													X	X	X	X	
Tomlinson	R													X	X	X		
Townsend	B																X	X
Vogt	M															X	X	X
Hurley	K															X	X	
Jobst	M															X	X	
Kirkman	R																X	
Hayslip	A																X	
Skoye	R																X	

C -= Chair; X = member; r = reviewer only; XC = Chair transition

Figure 9

GIS Management Institute Volunteers & Contributors 2008-2015

Acknowledgements

At the March 25, 2014 GMI Committee meeting David DiBiase suggested that the Geospatial Management Competency Model, the GIS Capability Maturity Model, the GMI Glossary, and an introductory editorial about the GIS Management Institute, could provide the content for a complete stand-alone issue of the *URISA Journal*. That suggestion was the original inspiration for this monograph. I appreciate review of the draft *URISA Journal* article and useful suggestions by A. Esnard, in particular the need for a chronology to help understand the various threads in the development of the GIS Management Institute. I am of course grateful to the dozens of GIS professionals who helped to develop the GIS Management Institute.

The opinions expressed and any errors in this work are mine alone.

About the Author

Greg Babinski, MA, GISP
Marketing & Business Development Manager, King County GIS Center
Past-President of URISA

Greg Babinski has been Finance and Marketing Manager for the King County GIS Center in Seattle, Washington since 1998. Before that he was the GIS Supervisor of the East Bay Municipal Utility District in Oakland, California. He holds BA and MA degrees in geography from Wayne State University in Detroit, Michigan. With 29 years of GIS experience, he is a certified GIS professional (GISP). As a member of the Government Finance Officers Association he has presented at GFOA conferences and written about GIS for Government Finance Review. He is also a member of the American Marketing Association (AMA), and of the International Association of Chinese Professionals in Geographic Information Sciences.

He has been active in the Urban and Regional Information Systems Association (URISA) and its chapters for more than 27 years. He served on the Board of the BAAMA Chapter and helped organize the California GIS Conference. He served on the Board of the Washington State Chapter and helped organize the annual Washington GIS Conference. He is also founder and former Chief Editor of *The Summit,* the Washington GIS Newsletter. He is a URISA workshop author and certified instructor. He helped author the URISA GIS Leadership Academy (ULA) and is a former ULA instructor. He served on URISA's Board of Directors for six years and was URISA President in 2011-2012.

He has written numerous articles about his interest and original research into geospatial industry management, marketing, finance, and budgeting. He was the author of the original GIS Capability Maturity Model and coordinated its review and revision for adoption by URISA in 2013. He is the founder and was the first chair of URISA's GIS Management Institute Committee. Greg has received the URISA Leadership Award and the WAURISA Summit Award and was named a 'GIS Hero' in the Winter 2012/2013 issue of *ArcNews.* Greg has spoken about GIS at many conferences and events in the US and Canada, as well as in the Caribbean, Australia, and across Europe and Asia.

In his spare time, Greg likes hiking steep narrow dangerous trails that lead high above the clouds to awesome views.

References

Albrecht, J. 2015. Towards a Theory of GIS Program Management. In Kuhn W and H Onsrud (Eds), Advancing Geographic Information Science: the past and the next 20 years. Needham, MA: GSDI Association Press, pp. 9-10.

Almvig, G. and Burdick, D. 2014. The Secret Mix of Successful GIS Project Management. 2014 Washington GIS Conference. http://www.waurisa.org/conferences/2014/presentations/The%20Secret%20Mix%20of%20Successful%20GIS%20Project%20Management.pdf.

Alrwais, O. A. 2016. A New GIS Maturity Model: An Organizational Usage Perspective. CGU Theses and Dissertations, Claremont Graduate University. 2016.

Alrwais, O., Horan, T., Hilton, B., and Bechor, T. 2015. Evaluating Local Government Usage of GIS: A New Maturity Model, Pre-ICIS Workshop on Location Analysis and Big Data, Fort Worth, 2015, see: http://www.redlands.edu/globalassets/depts/school-of-business/gisab/workshops-conferences/omer-alrwais-et-al_evaluating-local-government-usage-of-gis_pre-icis-workshop_12-3-15.pdf.

Anderson, C. 2013. Introducing URISA Labs. The *GIS Professional*, issue 252, January-February 2013, pp. 18-19.

Anderson, C. 2014a. My URISA Story. The *GIS Professional*, Issue 262, September-October 2014.

Anderson, C. 2014b. How Well Does Your GIS Measure Up?. The *GIS Professional*, Issue 263, November-December 2014.

Ayan, D. and Ouimet, M. 2011 NSGIC Geospatial Maturity Assessment: https://www.nsgic.org/gma-work-group.

Aziz, E. E. (2015). Project closing: the small process group with big impact. Paper presented at PMI® Global Congress 2015—EMEA, London, England. Newtown Square, PA: Project Management Institute.

Babinski, G. 1998. 1998 Mid-Year Survey of GIS User Department's Key Issues and Requirements. King County Coordinated Geographic Information System, September 24, 1998.

Babinski, G. 2002. GIS Knowledge Sharing: Using the Web for Open Business to Promote Regional GIS Success, 2002 URISA Annual Conference, see: http://www.slideshare.net/gbabinski/gis-knowledge-sharing-using-the-web-for-open-business-to-promote-regional-gis-success.

Babinski, G. 2003a. A Business Line Approach to Enterprise GIS Finance, 2003 URISA Annual Conference, see: http://www.slideshare.net/gbabinski/urisa2003papergreg-b?qid=bc7a5ed9-ee0d-4e19-9840-d693809c0760&v=&b=&from_search=9

Babinski, G. 2003b. 2002 KCGIS O&M Issue Status Report #4: Status of Original GIS Capital Project Work Tasks, King County GIS Technical Committee, August 26, 2003, see: https://www.slideshare.net/gbabinski/2002-kcgis-om-issue-status-report-4-status-of-original-gis-capital-project-work-tasks.

Babinski, G. 2004a. Effective Marketing: a Key Success Factor for Enterprise GIS, 2004 URISA Annual Conference, see: http://www.slideshare.net/gbabinski/effective-marketing-a-key-success-factor-for-enterprise-gis-2004-urisa-annual-conference.

Babinski, G. 2004b. Analyzing GIS Staff Usage and Allocations for Improved GIS Operations. 2004 Washington URUISA Annual Conference. See: http://www.slideshare.net/gbabinski/analyzing-gis-staff-usage-and-allocation-for-improved-gis-operations.

Babinski, G. 2005. Defining the Cost and Determining the Value of Enterprise GIS, 2005 URISA Annual Conference, see: http://www.slideshare.net/gbabinski/defining-the-cost-and-determining-the-value-of-enterprise-gis?qid=bc7a5ed9-ee0d-4e19-9840-d693809c0760&v=&b=&from_search=8.

Babinski, G. 2006. Entrepreneurial GIS Services: Innovative Practices at King County GIS, 2006 URISA Annual Conference, see: http://www.slideshare.net/gbabinski/entrepreneurial-gis-services-innovative-practices-at-king-county-gis-2006-urisa-annual-conference.

Babinski, G. 2007. Measuring Enterprise GIS Performance For the King County GIS Center, 2007 Washington URUISA Annual Conference, see: http://www.slideshare.net/gbabinski/measuring-enterprise-gis-performance-for-the-king-county-gis-center.

Babinski, G. 2008. Tracking the Anonymous GIS User in King County, 2008 URISA Annual Conference. http://www.slideshare.net/gbabinski/tracking-the-anonymous-gis-user-in-king-county?qid=bc7a5ed9-ee0d-4e19-9840-d693809c0760&v=&b=&from_search=12

Babinski, G. 2009a. A Municipal GIS Capability Maturity Model, unpublished working paper. See: https://www.slideshare.net/gbabinski/draft-municipal-gis-capability-maturity-model.

Babinski, G. 2009b. A Proposed Municipal GIS Capability Maturity Model. 2009 Washington URISA Conference. See: http://www.slideshare.net/gbabinski/a-proposed-municipal-gis-capability-maturity-model.

Babinski, G. 2009c. Draft Municipal GIS Capability Maturity Model. See: http://www.slideshare.net/gbabinski/municipal-gis-capability-maturity-model-questionnaire.

Babinski, G. 2009d. A Proposed GIS Capability Maturity Model. 2009 URISA Annual Conference. See: http://www.slideshare.net/gbabinski/gis-capability-maturity-model.

Babinski, G. 2010a. A Proposed GIS Capability Maturity Model. 2010 GIS in Action Conference.

Babinski, G. 2010b. Update on the Proposed Municipal GIS Capability Maturity Model. 2010 Washington GIS Conference.

Babinski, G. 2010c. An American Discovers Taiwan's Vibrant GIS Environment. The Summit, Autumn 2010. See: http://www.waurisa.org/thesummit/TheSummit_Autumn_2010.pdf.

Babinski, G. 2010d. Briefing: The URISA Draft Municipal GIS Capability Maturity Model, At National Geospatial Advisory Committee Meeting, Washington, DC, September 22, 2010, see: http://www.slideshare.net/gbabinski/gis-capability-maturity-model-briefing.

Babinski, G. 2010e. See: URISA Proposes a Local Government GIS Capability Maturity Model, Winter 201/2011 *ArcNews*. See http://www.esri.com/news/arcnews/winter1011articles/urisa-proposes.html

Babinski, G. 2011a. URISA GIS Capability Maturity Model. 2011 North Carolina GIS Conference Proceedings, 2011 (http://www.cgia.state.nc.us/ncgis2011/proceedings.aspx).

Babinski, G 2011b. You Can't Manage What you Don't Measure - URISA's Proposed Municipal GIS Capability Maturity Model, Keynote address at 2011 Alaska Surveying and Mapping Conference, Anchorage, Alaska, February 22, 2011, see: http://www.slideshare.net/gbabinski/you-cant-manage-what-you-dont-measure-urisas-proposed-municipal-gis-capability-maturity-model.

Babinski, G 2011c. Law and the Geoweb' Panel Discussion. 2011 AAG Conference. See: http://www.slideshare.net/gbabinski/law-and-the-geoweb-panel-discussion-2011-aag-conference.

Babinski, G. 2011d. URISA's Local Government GIS Capability Maturity Model, URISA Connect Webinar, August 10, 2011, see: http://www.slideshare.net/gbabinski/urisas-local-government-gis-capability-maturity-model.

Babinski, G. 2011e. The Future of URISA as an International Organization. Report to the URISA Board of Directors. December 19, 2011. See: https://www.slideshare.net/gbabinski/the-future-of-urisa-as-an-international-organization.

Babinski, G. 2012a. URISA Develops the Geospatial Management Competency Model for DOLETA. 2012 GIS in Action Conference. Portland, OR. 2012. See: http://www.slideshare.net/gbabinski/urisa-develops-the-geospatial-management-competency-model-gmcm-for-usdoleta.

Babinski, G. 2012b. URISA and the Development of a Geospatial Society, Keynote address at the inaugural event of the URISA-UAE chapter in Dubai, UAE, Towards Geospatial Society. See: http://www.slideshare.net/gbabinski/urisa-and-the-development-of-a-geospatial-society-past-present-and-future.

Babinski, G. 2012c. URISA Develops the Geospatial Management Competency Model (GMCM) for USDOLETA. Paper delivered at the 2012 Washington GIS Conference in Tacoma, WA. See: http://www.slideshare.net/gbabinski/urisa-develops-the-geospatial-management-competency-model-gmcm-for-usdoleta-60351773.

Babinski, G. 2012d. URISA's GIS Management Institute. Paper delivered at the 2012 Esri User Conference, San Diego, 2012. See: http://www.slideshare.net/gbabinski/urisas-gis-management-institute.

Babinski, G. 2012e. URISA's GIS Management Institute. Presentation for the Singapore Land Authority, August 6, 2012. See: http://www.slideshare.net/gbabinski/urisas-gis-management-institute-60357860.

Babinski, G. 2012f. URISA's GIS Management Institute. Presentation for the Singapore Land Authority, August 6, 2012. See: https://www.slideshare.net/gbabinski/urisas-gis-management-institute-88551748.

Babinski, G. 2012g. 'Geospatial Society, the GIS Profession, and URISA's GIS Management Institute' *ArcNews*, Autumn 2012, p. 43: http://www.esri.com/news/arcnews/fall12articles/geospatial-society-the-gis-profession-and-urisas-gis-management-institute.html.

Babinski, G. 2012h. URISA's 50 Years as an International Organization, *BURISA News*, Issue 192 (June 2012) pp. 12-13, Issue 193 (September 2012) pp. 12-14, Issue 194 (December 2012) pp. 12-15.

Babinski, G. 2012i. Reflections on URISA at 75, The *GIS Professional*, Issue 250. September-October 2012, pp. 3-7.

Babinski, G. 2012j. URISA Develops the Geospatial Management Competency Model (GMCM) for USDOLETA, Sixth URISA Caribbean Conference in Montego Bay, Jamaica, November 2012: http://www.slideshare.net/gbabinski/urisa-develops-the-geospatial-management-competency-model-gmcm-for-usdoleta-60521612.

Babinski, G. 2012k. URISA and the Development of a Geospatial Society: Past, Present and Future, Tsinghua University's Center for Earth Sciences, Redbud Forum on Global Change Science, December 5, 2012: http://www.slideshare.net/gbabinski/urisa-and-the-development-of-a-geospatial-society-past-present-and-future-60522015.

Babinski G. 2013a. 18 Steps to Success as a GIS Manager: URISA's Geospatial Management Competency Model, Keynote address at 2013 North Carolina GIS Conference in Raleigh, NC. See http://www.slideshare.net/gbabinski/18-steps-to-success-as-a-gis-manager-urisas-geospatial-management-competency-model.

Babinski G. 2013b. URISA, The Development of a Geospatial Society and Why GIS Matters, March 11, 2013, Keynote address at Northern Rockies URISA 2013 Intermountain GIS Conference in Boise, ID. See: http://www.slideshare.net/gbabinski/urisa-the-development-of-a-geospatial-society-and-why-gis-matters.

Babinski G. 2013c. URISA and the Development of a Geospatial Society, April 15-19, 2013, SSSI Conference, Canberra, Australia, see: http://www.slideshare.net/gbabinski/urisa-and-the-development-of-a-geospatial-society.

Babinski G. 2013d. URISA, The Development of a Geospatial Society, ROI, and Why GIS Matters, Imagin Conference, Mount Pleasant, MI, April 29, 2013, see: http://www.slideshare.net/gbabinski/urisa-the-development-of-a-geospatial-society-roi-and-why-gis-matters.

Babinski G. 2013e. URISA and the GIS Management Institute, 2013 Washington GIS Conference, Lynnwood, WA, May 7, 2013, see: http://www.slideshare.net/gbabinski/urisa-and-the-gis-management-institute.

Babinski, G. 2013f. Using ROI Methodology to Document the Benefits of King County GIS, Geospatial World Forum, Rotterdam, The Netherlands, May 12, 2013, see: http://www.slideshare.net/gbabinski/using-roi-methodology-to-document-the-benefits-of-king-county-gis.

Babinski, G. 2013g. The Role of Regional Geospatial Coordination for Effective City Management, Geospatial World Forum, Rotterdam, The Netherlands, May 13, 2013, see: http://www.slideshare.net/gbabinski/the-role-of-regional-geospatial-coordination-for-effective-city-management.

Babinski, G. 2013h. Best Practices for Financial Management of GIS Operations, September 17, 2013, GIS-Pro 2013, Providence, RI, see: http://www.slideshare.net/gbabinski/best-practices-for-financial-management-of-gis-operations.

Babinski, G. 2013i. URISA's GIS Management Institute® Updates the GIS Capability Maturity Model, September 17, 2013, GIS-Pro 2013, Providence, RI, see: http://www.slideshare.net/gbabinski/urisas-gis-management-institute-updates-the-gis-capability-maturity-model.

Babinski, G. 2013j. The GIS Management Institute and the GIS Capability Maturity Model (GISCMM): updates, overviews and implementations, URISA Connect Webinar recorded November 6, 2013. See: https://www.youtube.com/watch?v=QI33mQKj1mc&feature=youtu.be

Babinski, G. 2013k. GMI Business Plan Products and Services Matrix. Revised 28 August 2013.

Babinski, G. 2013l. URISA's International Affiliation with SSSI Continues – 2013 Trip report, The *GIS Professional*, Issue 254. May-June 2013, pp. 6-9.

Babinski, G. 2014a. Maximizing Benefits from Municipal GIS Operations The GIS Management Institute® and the GIS Capability Maturity Model, Lecture Delivered at South Ural State University, Chelyabinsk, Russia, January 21-22, 2014. See: http://www.slideshare.net/gbabinski/maximizing-benefits-from-municipal-gis-operations-the-gis-management-institute-and-the-gis-capability-maturity-model.

Babinski, G. 2014b. The GIS Management Institute® and the GIS Capability Maturity Model Update, Overview, and Implementation, Lecture delivered to the Southeast Washington & Northeast Oregon GIS User Group on April 3, 2014 in Walla Walla, WA. See: http://www.slideshare.net/gbabinski/the-gis-management-institute-and-the-gis-capability-maturity-modelupdate-overview-and-implementation.

Babinski, G. 2014c. URISA GIS Management Institute® The GIS Capability Maturity Model and the GMI® GIS Assessment Service. Paper delivered at the 2014 Esri User Conference, San Diego, July 16, 2014. See: http://www.slideshare.net/gbabinski/urisa-gis-management-institute-the-gis-capability-maturity-model-and-the-gmi-gis-assessment-service.

Babinski, G. 2014d URISA, the Geospatial Revolution, and the GIS Management Institute, Keynote address at 2014 New England URISA Day, Old Sturbridge Village, MA, November 3, 2014, see: http://www.slideshare.net/gbabinski/urisa-the-geospatial-revolution-and-the-gis-management-institute.

Babinski, G. 2015a. The State of GIS in Washington & Oregon The 2014 GMI Metric Survey, Paper delivered May 6, 2015 at 2015 GIS in Action Conference, Vancouver, WA, see: http://www.slideshare.net/gbabinski/the-state-of-gis-in-washington-oregon-the-2014-gmi-metric-survey.

Babinski, G. 2015b. Prospecting for Profits with GIS, presentation to the TDWI June 2015 meeting, Seattle, WA, see: http://www.slideshare.net/gbabinski/prospecting-for-profits-with-gis.

Babinski, G. 2015c. From the Emerald City to Your City GIS for Effective Service and ROI, 2015 ICMA Conference, Seattle, WA, September 29, 2015, see: http://www.slideshare.net/gbabinski/from-the-emerald-city-to-your-city-gis-for-effective-service-and-roi.

Babinski, G. 2015d. GIS Return on Investment, Webinar for the Federal Highway Administration GIS in Transportation, November 5, 2015, see: https://www.gis.fhwa.dot.gov/webcast27_id_wa.asp.

Babinski, G. 2016. Managing a GIS for Results, Paper presented at the American Society of Public Administration Conference in Seattle, WA on March 22, 2016, see:

Babinski, G. and Esnard, A., 2014. The GIS Capability Maturity Model: Maximize Benefits from Enterprise GIS Operations. URISA Connect Webinar, September 17, 2014. See: http://www.slideshare.net/gbabinski/the-gis-capability-maturity-model-maximize-benefits-from-enterprise-gis-operations-the-gis-management-institute.

Babinski, G. and Esnard, A. 2015a Building Sustainable GIS Management with URISA's GIS Capability Maturity Model, *ArcNews*, Spring, 2015, pp. 30-31, see: http://www.esri.com/esri-news/arcnews/spring15articles/building-sustainable-gis-manageement-with-urisas-gis-capability-maturity-model.

Babinski, G. and Esnard, A. 2015b. Update on the URISA GIS Management Institute GIS Assessment Service, paper presentation during GIS-Pro 2015 in Spokane, WA, October 21, 2015, see: http://www.slideshare.net/gbabinski/update-on-the-urisa-gis-management-institute-gis-assessment-service.

Babinski, G., Ibaugh, A, and Esnard, A. 2014. GIS Management Institute Update, plenary session presentation during GIS-Pro 2014 in New Orleans, September 10, 2014.

Babinski, G., Brown, C., Butler, A., Smith, C., DiBiase, D., Obermeyer, N., and Nelson, W. 2011. Proposal to Establish the GIS Management Institute. URISA, draft August 15, 2011 (revised 9/7/2011, 10/19/2011, and 10/23/2011.

Babinski, G., Perkins, H., McDermott, T.D., Ayan, D. 2010. Municipal GIS Capability Maturity Model Workshop. 2010 URISA GIS-Pro Conference.

Babinski, G., Butler, A., and Ibaugh, A. 2014. GIS Capability Maturity Model, URISA Certified Workshop, 2014.

Babinski, G., Horning, G., Higgins, D., Hocking, G., Leathers, M., and Stought, G. 2017. King County GIS Center 2015 GIS Assessment Report, *URISA Journal*, Vol. 27, No. 2, pp. 21-35.

Bailey, M., Feher, K., Kavanagh, S. 2011. Improve Performance Using the GFOA's New Financial Management Assessment Tool. Government Finance Review, August 2011, pp. 18-22.

Bloom, N., Brynjolfsson, E., Foster, L., Jarmin, R., DSapoorta-Eksten, I, Van Reenen, J. 2013. Center for Economic Studies (CES), U.S. Census Bureau. CES 13-01. January 2013.

Bossler, Dr. J.D., Dr. Cowen, Dr. D.J., Geringer, J.E., Lambert, S.C., Moeller, J.J., Rust, T.D., and Welch, R.T. Report Card on the U.S. National Spatial Data Infrastructure - Compiled for the Coalition of Geospatial Organizations. February 6, 2015. See: http://cogo.pro/uploads/COGO-Report_Card_on_NSDI.pdf. P. 15.

Bradbard, D.A., Fuller, B.K., and Townsend, B.B. 2017. Geographic Information Systems in Selected Cities and Counties in the Southeast United States. *URISA Journal*. Vol. 27, No. 2. 2017, pp. 5-20.

Butler, A. 2009. Undated email from Al Butler addressed to Hilary Perkins and included with the URISA Board meeting packet for its September 29, 2009 face-to-face meeting.

Butler, A. 2012. Directing URISA into the Future: A Proposal for Discussion, The *GIS Professional*, Issue 249. July-August, 2012, pp. 1-9.

Butler, A. 2013a. URISA Starts the Task of Reorganizing for the Future, The *GIS Professional*, Issue 252. January-February, 2013, pp. 8-16.

Butler, A. 2013b. Building a Stronger Association through the Chapters, The *GIS Professional*, Issue 252. January-February, 2013, pp. 20-21.

Butler, A. 2013c. What is Our Profession? The *GIS Professional*, Issue 253. March-April, 2013, pp. 7-10.

Butler, A. 2013d. President's Column, The *GIS Professional*, Issue 254. May-June, 2013, pp. 12-13.

Butler, A. 2013e. Migration of Historical URISA Operations to the New Structure, The *GIS Professional*, Issue 255. July-August, 2013, pp. 10-13.

Butler, A. 2013f. Welcome. Conference Program, GIS-Pro 2013: URISA's 51st Annual Conference for GIS Professionals. 2013. p. 2.

Butler, A. 2013g. Liability Disclaimer and Release for GIS Management Institute Subscription Services. URISA GIS Management Institute, 2013.

Butler, A., Ibaugh, A., and Babinski, G. 2014. GIS Capability Maturity. Workshop presented at GIS-T Symposium, Burlington, VT, May 5, 2014. See: http://www.gis-t.org/workshops/#wrk7.

City of Charlotte 2009. Best Practices for Enterprise GIS. Office of the CIO, City of Charlotte, North Carolina. Final Report, May 28, 2009.

Corbin, T. 2016. Ready for 2017. The *GIS Professional*, Issue 275, November-December 2016, pp. 14-15.

Corbin, T. 2017. Education and Activity. The *GIS Professional*, Issue 276, January-February 2017, pp. 14-15.

Croswell, P. 2015. Report on National Survey of Multi-Organization GIS Programs. GIS Management Institute Discussion Paper No. 2, February 2015.

DiBiase, D. 2012. Accreditation. Report to URISA GMI Committee, December 17, 2012.

DiBiase, D., DeMers, M., Johnson, A., Kemp, K., Luck, A.T., Plewe, B., and Wentz, E. 2006. Geographic Information Science and Technology Body of Knowledge. Washington, DC: Association of American Geographers.

DiBiase, D., Corbin, T., Fox, T., Francica, J., Green, K., Jackson, J., Jeffries, G., Jones, Brian, Jones, Brent, Mennis, J., Schuckman, K., Smith, C., and Van Sickle, J. 2010. The New Geospatial Technology Competency Model: Bringing Workforce Needs into Focus,' *URISA Journal*, Vol. 2, No. 2, 2010, pp. 55-72.

Director of National Intelligence. 2014. Information Sharing Environment, Geospatial Interoperability Reference Architecture (GIRA). Office of the Director of National Intelligence, 2014. See: https://www.dni.gov/files/ISE/documents/DocumentLibrary/GIRA.pdf.

Eklund, D., Richardson, H.M., Fletcher, M., McDaniel, N., and Sandler, R. 1997. Management Audit: Information and Telecommunication Services. Report 97-10 presented to the Metropolitan King County council by the County Auditor's Office. November 1997.

Esnard, A. 2014. Welcome. Conference Program of URISA's GIS-Pro 2014 Conference, 2014, p. 3.

Esnard, A., Ibaugh, A., Townsend, B. 2016. The GMI Road to Improved GIS Capabilities for Successful GIS Managers. URISA GIS-Pro Conference, 2016.

Finkle, R., Glowitz, J., and Mizula, L. 1997. East Bay Municipal Utility District CAD/CAM Best Practices Study. Plangraphics, Inc., April 18, 1997.

Fox, A. (Project Manager). 2018. GIS Process Rationalization Recommendation Report for King County Information Technology. The Timmons Group, Richmond, Virginia, February 2018.

Gaudet, C., H. Annulis, H.M., and Carr, J.C. 2003. Building the geospatial workforce. *URISA Journal* 15 (1): 21-30.

GIS Management Institute. 2012. Geospatial Management Competency Model. URISA. June 8, 2012. See: http://www.urisa.org/clientuploads/directory/GMI/Advocacy/GMCM%20final.pdf.

GIS Management Institute. 2013a. GIS Capability Maturity Model. URISA. September 2013. See: http://www.urisa.org/clientuploads/directory/GMI/GISCMM-Final201309(Endorsed%20for%20Publication).pdf.

GIS Management Institute. 2013b. GIS Management Institute Prospectus. Urban and Regional Information Systems Association. October 23, 2013.

GIS Management Institute. 2013c. The GIS Management Institute Publishes the URISA GIS Capability maturity Model, The *GIS Professional*, Issue 256. July-August, 2013, p. 8.

GIS Management Institute. 2014a. Glossary of Terms Used in the GIS Capability Maturity Model. URISA. May 8, 2014. See: http://www.urisa.org/clientuploads/directory/GMI/GISCMM_Glossary_5-8-14_Final.pdf

GIS Management Institute. 2014b. GIS Benchmarking Metrics Survey. URISA GIS Management Institute, May 29, 2014.

GIS Management Institute. 2014c. 2014 Product and Services Roll-out Scope of Work. GIS Management Institute Committee, 2014.

GIS Management Institute. 2015. Geospatial Management Competency Model Proficiency Rating Scale. URISA GIS Management Institute. June 2015.

Grams, S. 2004. The History of the GISCI Certification Program. Published on GISCI website, see: See: https://www.gisci.org/AboutUs/History.aspx.

Grand Rapids Executive Commission. 2012. Fourth Annual Information Technology and Change Management Report, City of Grand Rapids, February 3, 2012.

Green, M., Lucivero, A., Reovan, A., and Quinton, D. 2017. GIS in Transportation – Geospatial Information System Capability Maturity Models: Case Studies of Select Transportation Agencies, 2017. Federal Highway Administration. June 2017.

Hayes, T. 2014. GISCMM Assessment and Questionnaire Supporting Documentation. San Jose/Santa Clara Regional Wastewater Facility, San Jose, California, 2014.

Hitt, A. and Townsend, B. 2017a. URISA's GIS Management Institute: What Can GMI Do For Me? The *GIS Professional*, Issue 277, March-April 2017, p. 4.

Hitt, A. and Townsend, B. 2017b. Getting started with the GISCMM in 3 quick steps. The *GIS Professional*, Issue 278, May-June 2017, p. 4.

Hitt, A. and Townsend, B. 2017c. What Happens After the Assessment? The *GIS Professional*, Issue 279, July-August 2017, p. 6.

Hitt, A. and Townsend, B. 2017d. URISA's GIS Management Institute: Bringing It All Together. The *GIS Professional*, Issue 280, May-June 2017, pp. 1-2.

Ibaugh, A. 2013a. From Student Member to URISA President. The *GIS Professional*, Issue 256. September-October, 2013, pp. 10-13.

Ibaugh, A. 2013b. "Urban and Regional" Information Systems Association – 50+ Years Later, and Still More Relevant Than Ever. The *GIS Professional*, Issue 257. November-December, 2013, pp. 13-14.

Ibaugh, A. 2014a. Ibaugh, Allen, What's in a Tag Line? – "Fostering Excellence in GIS". The *GIS Professional*, Issue 258, January-February 2014, p. 8.

Ibaugh, A. 2014b. Ibaugh, Allen, GIS Management Institute: What is it? The *GIS Professional*, Issue 259, March-April 2014.

Ibaugh, A. 2014c. Ibaugh, Allen, GIS-Pro 2014 – Leveraging Spatial Technology to Support Sustainable and Resilient Communities. The *GIS Professional*, Issue 260, May-June 2014.

Ibaugh, A. 2014d. Ibaugh, Allen, President's Farewell Letter. The *GIS Professional*, Issue 261, July-August 2014.

Ibaugh, A. 2017. GIS Capability Maturity Modeling at State DOTs. URISA GIS-Pro Conference Paper, 2017.

IndraStar 2016. A View on GIS Capability Maturity Model. April 11, 2016, see: http://www.indrastra.com/2016/04/IT-GIS-Capability-Maturity-Model-002-04-2016-0030.html.

Karalopoulus, A. and Kavouras, M. 2015 Survey on Maturity Models for the Geospatial Domain, 18th Agile International Conference of Geographic Information Science, Lisbon, June 9-12, 2015.

Makela, J, 2012. Model for Assessing GIS Maturity of an Organization, Aalto University, Helsinki, 2012.

Mangan, M. 2008. Introducing a Maturity Model for Enterprise GIS. Even Keel Strategies, Inc. Chicago. 2008.

McLeod, B. 2014. Using the URISA GIS Capability Maturity Model. The Gartrell Group, 2014.

National States Geographic Information Council. 2015. 2015 Geospatial Maturity Assessment. NSGIC. 2015. See:

Niessink, F., Clerc, V., Tijdink, T., and van Vliet, H. 2005. The IT Service Capability Maturity Model. CIBIT Consultants, Amsterdam, The Netherlands, January 28, 2005.

Obermeyer, N. 1993. Certifying *GIS Professionals*: Challenges and Alternatives, *URISA Journal,* Vol. 5, No. 1, Spring, 1993, pp. 67-75.

Office of Internal Audit 2012. Performance Audit of the Geographic Information System (GIS) Final Report. Parks Canada Agency, July, 2012.

Onsrud, H. and Kuhn, W. 2016. Advancing Geographic Information Science: The Past and Next Twenty Years. GSDI Association Press, 2016.

Peloquin, W. and Bowe, R. (2015). Charting the Path Forward: URISA Directions for GIS Best Practices and Educational Development, The *GIS Professional,* Issue 265, March-April 2015, pp. 7-8.

Price, M. 2015. Is local government GIS supporting the vision of the NSDI? The *GIS Professional,* Issue 266, May-June 2015, pp. 1-4.

Sarmiento, M, Green, M., and Lucivero, A. 2016. GIS Capability Maturity Model Peer Exchange, 2016. Nashville, TN, Federal Highway Administration. September 20-21, 2016.

Smith, C. 2010 President's Column. The *GIS Professional.* Issue 240, November-December 2010, pp. 8-9.

Smith, C. 2011a President's Column: URISA's Value. *The* GIS Professional. Issue 242, March-April 2011, pp. 5-6.

Smith, C. 2011b President's Column: Data Sharing Solution – Part 2. The *GIS Professional.* Issue 245, September-October 2011, pp. 8, 11.

Smith, C. and Ayan, D. 2014. GIS and the Three Pillars of Sustainability. The *GIS Professional,* Issue 262, September-October 2014, pp. 1-2.

Somers, R. 2015. URISA Rolls in 2016. The *GIS Professional,* Issue 269, November-December 2015, p. 12.

Somers, R. 2016a. GIS Program Review and Assessment. The *GIS Professional,* Issue 271, March-April 2016, pp. 16-17.

Somers, R. 2016b. URISA Continues to Grow the GIS Profession. The *GIS Professional,* Issue 272, May-June 2016, p. 10.

Somers, R. 2016c. Managing GIS by Best Practices. The *GIS Professional,* Issue 273, July-August 2016, pp. 10-11.

Somers, R. 2016d. Help URISA Grow the GIS Profession. The *GIS Professional,* Issue 274, September-October 2016, p. 12.

Somers, R. 2017. GIS Program Review, Assessment, and Improvement. URISA GIS-Pro Conference, 2017.

Sonnen, D. Moeller, J, and LaBranche, D. 2009. Geospatial Enterprise Integration Maturity Model. ISSI, Bellingham, Washington, ISD 2009-0054. June 24, 2009 revision of White Paper originally published March 2009.

Sparks, J., Worall, P., and Mickey, K. (2014) A Distributed Model for Effective National Geospatial Data Management: Building a National Data Sharing Infrastructure. GIS Management Institute Discussion Paper No. 1, March, 2014.

Terry, G. 2011. GIS-Pro 2011. The *GIS Professional.* Issue 244, July-August 2011, pp. 1-3.

Townsend, B. 2017. Getting Started with the GISCMM, Local Government Experiences with GIS Program Assessment and Improvement. URISA GIS-Pro Conference, 2017.

Townsend, T. 2014. URISA Brain Trust – Committee Meetings at GIS-Pro 2014. The *GIS Professional,* Issue 262, September-October 2014, p. 11.

Townsend, T. 2017. Plan to Action. The *GIS Professional,* issue 281, November-December 2017, pp. 12-14.

URISA, 1986. URISA Constitution. The Urban and Regional Information Systems Association. Revised, Fall 1986.

URISA GMI Committee, Babinski, G. Chair, with Brown, C., Butler, A., Smith, C., DiBiase, D., Obermeyer, N., and Nelson, W., 2012. Proposal to Establish the GIS Management Institute, April 16, 2012.

URISA and ICA 2014. Memorandum of Understanding between URISA and the International Cartographic Association. 2014.

URISA and SSSI 2012. URISA GIS Capability Maturity Model and Geospatial Management Competency Model. SSI-URISA Joint Activity Agreement. June 2012.

URISA and SSSI 2013. 2013-2014 URISA GIS Management Institute. SSSI-URISA Joint Activity Agreement. 2013.

U.S. Department of Labor Employment Training Administration (USDOLETA) 2015. Competency Model Development and Use – A Technical Assistance Guide, 2015.

Warnecke, L., Beattie, J., Kollin, C., and Winifred, L. 1998. Geographic Information Technology in Cities and Counties: A Nationwide Assessment. Urban and Regional Information Systems Association and American Forests. 1998.

Washington URISA GISCMM Committee, Babinski, G., Beimburn, S., Burdick, D., Esnard, A., Griffin, T., Horning, G. and Von Essen, I. 2011. URISA Draft Municipal GIS Capability Maturity Model. May 2011. See: http://www.slideshare.net/gbabinski/urisa-giscmm-may-2011-draft

Washington URISA GMCM Committee, 2011. Babinski, G., Beimburn, S., Burdick, D., Esnard, A., Griffin, T., Horning, G. and Von Essen, I. 2011, Geospatial Management Competency Model Strawman Draft, September 10, 2011. See: https://www.slideshare.net/gbabinski/urisa-geospatial-management-competency-model-strawman-draft.

Wellar, B. (Ed.) 2012. Foundations of Urban and Regional Information Systems and Geographic Information Systems and Science. Urban and Regional Information Systems Association, Des Plaines, IL. 2012.

Zerbe, R., Fumia, D., Reynolds, T., Singh, P., Scott, T., and Babinski, G., 2016. An Analysis of Benefits from Use of Geographic Information Systems by King County, Washington, *URISA Journal,* Vol. 27, No. 1, pp. 13-28.

End Notes

---

[i] Jochen Albrecht email, April 12, 2017.

[ii] Jochen Albrecht email, June 10, 2017.

[iii] Cited in http://www.urisa.org/clientuploads/directory/Documents/Committees/URISA%20Committees2015.pdf, accessed March 10, 2016.

[iv] Cited in http://www.urisa.org/gmi, accessed November 5, 2017.

[v] The currently adopted URISA Bylaws were not available on the URISA website as of March 10, 2016.

[vi] https://en.wikipedia.org/wiki/Capability_Maturity_Model

[vii] I served on the URISA Board of Directors from 2007 until 2013.

[viii] http://www.careeronestop.org/competencymodel/competency-models/geospatial-technology.aspx. Not to be confused with the 2001 USM GTCM.

[ix] URISA Balanced Scorecard dated October 27, 2009.

[x] Wikipedia article: Institute, https://en.wikipedia.org/wiki/Institute, accessed March 6, 2018.

[xi] This informal proposal was made by URISA Past-President Barry Wellar.

[xii] URISA Board of Director Meeting Minutes, November 3, 2010.

[xiii] URISA Board of Directors Meeting Minutes, August 6-7, 2011

[xiv] Email to URISA Board of Directors, September 11, 2011.

[xv] URISA GIS Management Institute Committee Activity Report, dated October 23, 2011

[xvi] URISA Board of Directors Meeting Minutes, October 31, 2011

[xvii] URISA Board of Directors Meeting Minutes, November 4, 2011

[xviii] For the full published URISA Geospatial Management Competency Model, see: http://www.urisa.org/clientuploads/directory/GMI/Advocacy/GMCM%20final.pdf. (URISA would not permit me to publish the full Geospatial Management Competency Model, in this monograph).

[xix] GeoTech Center: Geospatial Technology Competency Model. See: http://www.careeronestop.org/competencymodel/competency-models/geospatial-technology.aspx.

[xx] URISA Board of Directors Meeting Minutes, September 15, 2013

[xxi] For the full published URISA GIS Capability Maturity Model, see: http://www.urisa.org/clientuploads/directory/GMI/GISCMM-Final201309(Endorsed%20for%20Publication).pdf. (URISA would not permit me to publish the full GIS Capability Maturity Model in this monograph).

[xxii] URISA Board of Directors Meeting Minutes, February 24, 2012

[xxiii] Sheila Wilson Email 6/8/2012.

[xxiv] Mike Van Hook email, June 18, 2012.

[xxv] I developed the GMI hex design to represent the GMI concept in early 2012 and first presented it publically during the GMI launch announcement at the 2012 Esri User Conference.

[xxvi] URISA Annonces GIS Management Initiative, URISA, Des Plaines, IL, July 25, 2012. See: http://www10.giscafe.com/nbc/articles/view_article.php?articleid=1107671

[xxvii] GMI Committee Meeting Minutes, October 2, 2012.

[xxviii] URISA Board of Director Meeting Minutes, January 3, 2013.
[xxix] URISA Board of Director Meeting Minutes, March15-16, 2013.
[xxx] URISA Bylaws Revised March 16, 2013.

[xxxi] See: http://www.gfoa.org/.

[xxxii] See: http://www.cipfa.org/services/advisory-and-consultancy/financial-management-model.

[xxxiii] Data Management Association International (DAMA), see: www.dama.org).
[xxxiv] From DAMA website (www.dama.org) accessed March 22, 2018.
[xxxv] URISA Proposal to Establish the GIS Management Institute, Revised September 7, 2011

[xxxvi] URISA GIS Management Institute Business Plan, August 23, 2013.

[xxxvii] URISA GIS Management Institute, Business Plan: Financial Plan, Draft 28 August 2013.

[xxxviii] Inter-American Development Bank (IADB) see: https://www.iadb.org/en.
[xxxix] This contact was facilitated by Andrea Huber, Esri Global Affairs Account Manager. URISA worked with Sebastian Lew, Urban Planning Analyst, Emerging and Sustainable Cities Initiative, Inter-American Development Bank.
[xl] GMI Committee Meeting Minutes, February 25, 2014.

[xli] Thanks to Allen Ibaugh, URISA Past-President and CEO of DTS LLC.

[xlii] Office of Human resources at the National Institute of Health, Competencies Proficiency Scale, 2009, see: https://hr.od.nih.gov/workingatnih/competencies/proficiencyscale.htm.

[xliii] GMI Committee Meeting Minutes, January 27, 2015.

[xliv] Key International Cartographic Association contacts included Georg Gartner and Marcus Jobst. I made initial contacts with them during URISA events in Jamaica (2012) and St. Marteen (2014). Amy Esnard later facilitated meetings in Europe and championed the URISA-ICA MOU.
[xlv] Babinski email to Carl Anderson and selected URISA Board members, November 4, 2014. URISA cooperation with South Ural State University (SUSU) began in 2012 when Bill Huxhold conducted a URISA GISCorps mission to lecture on cadastral mapping practices. I later contacted SUSU at the 2013 Geospatial World Forum in Rotterdam, and again during my 2014 lecture at SUSU in Chelyabinsk in early 2014. Shortly after the preliminary SUSU-URISA discussion at the New England URISA event in November 2014, Russia annexed Crimea. After the U.S. imposed sanctions on Russia, there have been no further contacts from SUSU.
[xlvi] See: http://www.infrastructurereportcard.org/.

[xlvii] My DHS Contact was Ron Langhelm, First Responders Group, Science and Technology, U.S. Department of Homeland Security.
[xlviii] I proposed this GMI Knowledge Base at GIS-Pro 2015.
[xlix] My membership on this committee is documented in the minutes of the October 16, 2014 URISA Board of Directors meeting. Other Core Documents Committee members were Glenn O'Grady, Dianne Haley, Ed Wells, and Chair Kevin Mickey.
[l] I was also informed prior to consideration of my alternate proposal that I was no longer a voting member of the URISA Core Documents Committee by the Committee Chair.

[li] Email from Kevin Mickey to Core Documents Committee July 13, 2015.

[lii] Babinski email to R Somers, C. Anderson, A. Ibaugh, and T Corbin, November 10, 2015.

[liii] *URISA Adopts New and Improved Organizational Structure*, The *GIS Professional*, Issue 268, September-October 2015, p. 10.

[liv] The URISA Policy Manual can now be found at: http://www.urisa.org/about-us/urisa-board-of-directors-2017-2018/

[lv] URISA Bylaws, Adopted October 18, 2015, see: http://www.urisa.org/clientuploads/directory/Documents/BoardReports/URISA%20Bylaws_Adopted10182015.pdf

[lvi] URISA Committees, undated document on URISA website, see: http://www.urisa.org/clientuploads/directory/Documents/Committees/URISA%20Committees2015.pdf.

[lvii] Minutes of the URISA Board of Directors Meeting of October 22, 2015.

[lviii] URISA website accessed January 18, 2018, www.urisa.org.

[lix] *URISA Adopts New and Improved Organizational Structure,* The *GIS Professional*, Issue 268, September-October 2015, p. 10. This article listed the members of the 'Core Documents Committee' but omitted my name.

[lx] *Abstract Submissions Invited for GIS-Pro 2017 in Jacksonville, Florida*, The *GIS Professional*, Issue 276, January-February 2017, pp.3-4.

[lxi] Corbin, T., *Presidents Column: GIS has sprung*, The *GIS Professional*, Issue 277, March-April 2017, p. 10.

[lxii] Corbin, T., *Presidents Column: No summer vacation for URISA*, The *GIS Professional*, Issue 278, May-June 2017, pp. 8-9.

[lxiii] Email to author from W. Nelson, December 7, 2017.

[lxiv] Corbin, T., *Presidents Column: Education, Advocacy & OneURISA*, The *GIS Professional*, Issue 279, July-August 2017, pp. 10-11.

[lxv] Corbin, T., Presidents Column: Education, Advocacy & OneURISA, The *GIS Professional*, Issue 280, September-October 2017, pp. 6-7.

[lxvi] URISA website www.urisa.org, accessed March 13, 2018.

[lxvii] Email from R. Somers, February 16, 2018.

[lxviii] URISA GMI Website at www.urisa.org/gmi, accessed March 1, 2018.

[lxix] From the GIS Code of Ethics.

Made in United States
Orlando, FL
07 November 2026

90337481R00069